TEST AUTOMATION

BCS, THE CHARTERED INSTITUTE FOR IT

BCS, The Chartered Institute for IT, is committed to making IT good for society. We use the power of our network to bring about positive, tangible change. We champion the global IT profession and the interests of individuals, engaged in that profession, for the benefit of all.

Exchanging IT expertise and knowledge
The Institute fosters links between experts from industry, academia and business to promote new thinking, education and knowledge sharing.

Supporting practitioners
Through continuing professional development and a series of respected IT qualifications, the Institute seeks to promote professional practice tuned to the demands of business. It provides practical support and information services to its members and volunteer communities around the world.

Setting standards and frameworks
The Institute collaborates with government, industry and relevant bodies to establish good working practices, codes of conduct, skills frameworks and common standards. It also offers a range of consultancy services to employers to help them adopt best practice.

Become a member
Over 70,000 people including students, teachers, professionals and practitioners enjoy the benefits of BCS membership. These include access to an international community, invitations to a roster of local and national events, career development tools and a quarterly thought-leadership magazine. Visit www.bcs.org/membership to find out more.

Further information
BCS, The Chartered Institute for IT,
3 Newbridge Square,
Swindon, SN1 1BY, United Kingdom.
T +44 (0) 1793 417 417
(Monday to Friday, 09:00 to 17:00 UK time)
www.bcs.org/contact
http://shop.bcs.org/

TEST AUTOMATION
A manager's guide

Boby Jose

Published by BCS Learning and Development Ltd, a wholly owned subsidiary of BCS, The Chartered Institute for IT, 3 Newbridge Square, Swindon, SN1 1BY, UK.
www.bcs.org

Paperback ISBN: 978-1-78017-5454
PDF ISBN: 978-1-78017-5461
ePUB ISBN: 978-1-78017-5478

Ebook available

British Cataloguing in Publication Data.
A CIP catalogue record for this book is available at the British Library.

Disclaimer:
The views expressed in this book are of the author and do not necessarily reflect the views of the Institute or BCS Learning and Development Ltd except where explicitly stated as such. Although every care has been taken by the authors and BCS Learning and Development Ltd in the preparation of the publication, no warranty is given by the authors or BCS Learning and Development Ltd as publisher as to the accuracy or completeness of the information contained within it and neither the authors nor BCS Learning and Development Ltd shall be responsible or liable for any loss or damage whatsoever arising by virtue of such information or any instructions or advice contained within this publication or by any of the aforementioned.

All URLs were correct at the time of publication.

Publisher's acknowledgements
Reviewers: Kari Kakkonen, Francisca Cano Ortiz and Matthew Riddiough Boylan
Publisher: Ian Borthwick
Commissioning editor: Rebecca Youé
Production manager: Florence Leroy
Project manager: Sunrise Setting Ltd
Copy-editor: Gillian Bourn
Proofreader: Barbara Eastman
Indexer: John Silvester
Cover design: Alex Wright
Cover image: iStock - Leonid Andronov
Typeset by Lapiz Digital Services, Chennai, India.

To all my mentors who have inspired and encouraged me

CONTENTS

LIST OF FIGURES AND TABLES

FIGURES

TABLES

AUTHOR

Boby Jose holds a BSc in Physics and Masters in Business Administration (MBA), and has over two decades of work experience in software testing within the business, technology, IT, infrastructure, outsourcing and consulting industries. He started his professional career as a business analyst and ecommerce consultant. Professionally trained and certified in test automation, Boby's career progressed through offshore IT service companies, cutting edge product development and testing in informatics, and test consulting with a leading North American technology consultancy, and later with a leading European IT consultancy. He has worked in companies ranging from 500 to 250,000 employees and has led testing engagements with more than 100 members.

Boby has successfully managed the testing of many large IT transformation programmes and globally distributed software engagements. He is based in London, United Kingdom, and has software testing experience in the public, private, secure and product sectors in the US, Europe, Middle East and Asia. Boby has extensive experience in the fields of test strategy, test management, test planning, Test automation, test governance, test transformation, test environment management, infrastructure test management, test consulting and product test management.

Boby also has substantial testing and test automation experience in agro, bioinformatics, client/server, custom and bespoke solutions, digital, ecommerce, education, enterprise applications, enterprise resource planning (ERP), finance, healthcare, legacy, life science, marketplace, material science, mobile, multichannel, retail, scientific, tax, telecom, travel and transport, and web technologies. His skills spread across Agile, continuous integration (CI) and continuous delivery (CD), functional testing, infrastructure testing, iterative, IT service continuity (ITSC) testing, non-functional testing, operational acceptance testing (cloud, on-premises and hybrid), performance testing, portfolio, project and programme test management, security and penetration testing, test strategies and plans, user acceptance testing (for over 100 countries), V-model and Waterfall.

He is a certified ISTQB® Advanced Test Manager, as well as having qualifications that include TPI NEXT® Foundation, ISTQB-ISEB Foundation Level, PRINCE2® Foundation and Practitioner, Mobile App Testing, Certified Agile Scrum Master and Certified Scrum Product Owner.

He has published many papers, articles and points of view on topics such as the intricacies of mobile application testing, managing up in testing, process improvement through goal problem approach, investing in test automation (cost-benefit analysis and return on investment (ROI)) and software testing of product versus application-based approaches. In addition to this, Boby has previously managed testing for an 'award-winning mobile application' (Mobile Application of the Year by the Real IT forum).

You can find Boby on LinkedIn at 'Boby Jose' (https://www.linkedin.com/in/boby-jose-893b513/).

ABBREVIATIONS

AI	Artificial intelligence
ALM	Application life cycle management
API	Application programming interface
APO	Advanced planning and optimisation
ATDD	Acceptance test-driven development
ATF	Automated test framework
ATT	Automated testing tool
AUT	Application under test
AWS	Amazon Web Services
BAU	Business as usual
BDD	Behaviour-driven development
CapEx	Capital expenditure
CBA	Cost-benefit analysis
CC	Change control
CI	Continuous integration
CI/CD	Continuous integration and continuous delivery
CIT	Component integration testing
CMDB	Configuration management database
COTS	Commercial off-the-shelf
CRM	Customer relationship management
CSV	Comma separated value
CT	Component testing
CTO	Chief technology officer
DDD	Detailed design document
DDT	Defect detection trends
DL	Deep learning
DMT	Data migration testing
DOM	Document Object Model
DRY	Don't repeat yourself
E2E	End-to-end

ELS	Early life support
ERP	Enterprise resource planning
ESB	Enterprise service bus
ETL	Extract, transform, load
FAT	Factory acceptance testing
FOSS	Free and open-source software
FT/FP	Fixed time and fixed price
GDPR	General Data Protection Regulation
GIS	Geographic information system
GPL	General Public Licence
GPS	Global Positioning System
gTAA	Generic Test Automation Architecture
GUI	Graphical user interface
HIPAA	Health Insurance Portability and Accountability Act
HLD	High-level design
HR	Human resources
HTML	HyperText Markup Language
HTTP	HyperText Transfer Protocol
HTTPS	HyperText Transfer Protocol Secure
I18N	Internationalisation
IaaS	Infrastructure as a service
IAM	Identity and access management
IE	Internet Explorer
IP	Internet Protocol
ISCP	International supply chain planning
ISEB	Information System Examination Board
ISO	International Organization for Standardization
ISTQB	International Software Testing Qualifications Board
IT	Information technology
ITSC	IT service continuity
JAWS	Job Access With Speech
JD	Job description
JS	JavaScript
KPI	Key performance indicator
L10N	Localisation
LLD	Low-level design
LOE	Level of effort
ML	Machine learning
MRD	Manufacturing, retail and distribution

MS	Multi-supplier
MS	Microsoft
MTS	Managed testing services
MV	Multi-vendor
MVP	Minimum viable/valuable product
OAT	Operational acceptance testing
OEM	Original equipment manufacturer
OOP	Object-oriented programming
OpEx	Operational expenditure
OS	Operating system
PaaS	Platform as a service
Pen test	Penetration testing
PHP	Hypertext Preprocessor
PII	Personal identifiable information
PoC	Proof of concept
PoT	Proof of technology
PPR	Payroll parallel run
PVT	Performance and volume testing
Q&A	Question and answer
QA	Quality assurance
QTP	QuickTest Professional
RACI	Responsible, accountable, consulted and informed
RAG	Red, amber and green
RAID	Risks, assumptions, issues and dependencies
RBT	Risk-based testing
REST	Representational state transfer
ROI	Return on investment
RPA	Robotic process automation
RPO	Recovery point objective
RTM	Requirement traceability matrix
RTO	Recovery time objective
RUP	Rational Unified Process
SaaS	Software as a service
SDLC	Software development life cycle
SI	System integrator
SIT	System integration testing
SLA	Service level agreement
SME	Subject matter expert
SOA	Service-oriented architecture

SOAP	Simple Object Access Protocol
SQL	Structured Query Language
SSO	Single sign on
ST	System testing
STAF	Software test automation framework
STLC	Software test life cycle
SUT	System under test
T&M	Time and materials
TaaS	Testing as a service
TAS	Test automation solution
TDD	Test-driven development
TFS	Team Foundation Server
TMMi	Test Maturity Model integration
UAT	User acceptance testing
UFT	Unified Functional Testing
UI	User interface
UT	Unit testing
UX	User experience
VSTS	Visual Studio Team Services
VUs	Virtual users
W3C	World Wide Web Consortium
WCAG	Web Content Accessibility Guidelines
WMI	Windows Management Instrumentation
WORA	Write Once, Run Anywhere
WORE	Write Once, Run Everywhere
WSH	Windows Scripting Host
XML	Extensible Markup Language
XP	Extreme programming
XSS	Cross-site scripting

USEFUL WEBSITES

Accessibility testing	https://dynomapper.com/blog/27-accessibility-testing
Agile	https://blog.planview.com https://www.agilealliance.org/agile101/agile-glossary
Artificial intelligence	https://bernardmarr.com https://www.parasoft.com
Automation framework	https://bellatrix.solutions https://www.browserstack.com https://www.softwaretestinghelp.com https://www.toolsqa.com https://devops.com
Automation metrics	https://www.a1qa.com https://www.getzephyr.com https://blog.qatestlab.com
CI/CD	https://martinfowler.com/articles/continuousIntegration.html https://martinfowler.com/tags/extreme%20programming.html https://www.atlassian.com/continuous-delivery/continuous-integration/how-to-get-to-continuous-integration https://continuousdelivery.com https://stackify.com/continuous-delivery-vs-continuous-deployment-vs-continuous-integration https://bitbar.com/blog/top-continuous-integration-tools-for-devops https://apiumhub.com/tech-blog-barcelona/benefits-of-continuous-integration https://www.simplilearn.com/tutorials/devops-tutorial/continuous-integration https://aws.amazon.com/devops/continuous-integration https://martinfowler.com/articles/continuousIntegration.html https://www.agilealliance.org/agile101/agile-glossary https://dl.acm.org/doi/abs/10.1145/3236024.3275528

Certification	https://www.bcs.org https://techcanvass.com
Cloud solutions	https://phoenixnap.com
Coding standards	http://themoderndeveloper.com/code-standards
Development approaches	https://blog.planview.com
DevOps	https://devops.com
Encyclopedia	https://www.pcmag.com/encyclopedia/term/mts
Glossary	https://glossary.istqb.org/app/en/search
Mobile testing	https://www.mobindustry.net
Object-oriented programming	https://docs.oracle.com/javase/tutorial/java/concepts https://www.w3schools.com/cpp/cpp_oop.asp
Programming languages	https://www.codingninjas.com https://www.w3schools.com/cpp/cpp_oop.asp
Software testing	https://blog.qatestlab.com
Test automation	https://dzone.com https://www.capgemini.com https://www.codeproject.com https://www.katalon.com https://www.leapwork.com/blog/why-good-test-environments-are-crucial-for-successful-automation https://www.techwell.com https://www.testim.io/home-with-form
Test automation architecture	https://www.istqb.org/downloads/category/48-advanced-level-test-automation-engineer-documents.html https://labs.sogeti.com/benefits-of-generic-test-automation-architecture https://huddle.eurostarsoftwaretesting.com/test-automation-framework-architecture-types https://medium.com/koderlabs/introduction-to-monolithic-architecture-and-microservices-architecture-b211a5955c63
Test-driven design	https://www.madetech.com
Testing resources	http://www.softwareqatest.com https://www.istqb.org https://www.stickyminds.com https://bernardmarr.com

Testing skills	https://testsigma.com https://uk.indeed.com/?r=us
Testing standards	https://www.bcs.org/membership/ member-communities/software-testing-specialist-group https://softwaretestingstandard.org
Testing tools	https://briananderson2209.medium.com/ best-automation-testing-tools-for-2018-top-10-reviews- 8a4a19f664d2 https://docs.katalon.com/katalon-studio/docs/index.html https://reqtest.com https://www.altexsoft.com https://www.edureka.co https://www.knowledgehut.com https://www.methodsandtools.com https://www.saviantconsulting.com
Testing trends	https://techbeacon.com/app-dev-testing/5-ways-ai-will- change-software-testing
Testing types	https://geteasyqa.com/qa/software-testing-types
Training	https://testinginstitute.com/Company.php
Tutorial	https://www.guru99.com

PREFACE

Software testing today faces numerous challenges, including ever-shrinking release schedules, a lack of knowledge on business priorities, the misunderstanding of development methodologies, complex technologies, changing expectations, delays within implementation, a lack of awareness of non-functional requirements, insufficient resourcing, time and cost pressure versus quality, and the unfamiliarity of testing conventions. Most of these challenges become a risk and later a reality soon after the software solution development commences. Software products or projects try to mitigate the above issues in various ways, such as through planning and replanning, adding more people to the team or additional budget; however, testing is one of the key areas to be viewed as a mitigation to address the delay in delivery due to the above issues.

The testing team is often asked to reduce the test effort by following a risk-based approach, optimising testing stages, descoping testing types, merging testing phases and so on to compensate for the delay in the overall delivery schedule. Experienced and seasoned test managers are aware of these challenges and often plan at the beginning to address them. Test managers are often pushed against the wall to reduce the testing timeline in order to absorb delays in the planning, design and implementation phases. Test automation is one of the preferred ways of addressing these challenges by minimising the impact on overall quality.

This book is primarily written for test managers, Scrum masters, product and project managers on how to implement test automation and automated testing in their organisation.

Test automation is to automate the end-to-end testing process and activities, including tracking and managing the different tests. Automated testing is performing specific tests, such as performance testing and functional testing, via automation rather than manual means. (Refer to Chapter 1 for more details on test automation and automated testing.)

This book is a manager's guide to building and leading successful test automation at an organisational level.

Software testing, test automation and automated testing require more people who have good experience in the full project life cycle from the project concept to implementation, and early life support (ELS) to closure and decommission. Software testing demands good subject matter experts (SMEs) in end-to-end (E2E) testing, test automation, functional and non-functional testing, infrastructure testing, performance and volume testing, security testing, accessibility testing, recovery point objective (RPO) and recovery time objective (RTO) validation etc. Test organisations face challenges on the level of testing required, test coverage and expectations to complete testing as quickly and thoroughly as possible under the constraints of limited budget and time. Manual testing is labour-intensive, requires a good understanding of the application, knowledge of testing methodologies and skilful testers with good attitude and aptitude. While test automation complements manual testing, it requires detailed planning and support from the right stakeholders in the organisation. Testing tools and resources come at a high price, and test automation is effort-intensive with significant capital expenditure (CapEx). Due to the disparate capital expenditure involved, Test automation, if not implemented properly, will not provide the expected outcome while also increasing operational expenditure (OpEx). The goal of test automation is to automate the end-to-end testing process and activities, including tracking and managing the different tests. Automated testing is performing specific tests and so on, such as performance testing, functional testing and so on, via automation rather than manual means.

To implement automation tools, techniques and a framework, well thought through test strategy, test plan and test approach are required. There are numerous success stories of test automation and its benefits to showcase; however, this space has also recorded more failures than triumphs. The most common challenges in test automation are the lack of understanding of the system under test (SUT), choosing the right testing tools, and calculating effort and costs to blindly mimic other success stories with limited validation of your needs. Test automation frameworks often offer quick testing of large volumes of test cases and data validation alongside reporting capabilities that are mostly cost effective. Test automation can save resources and money in the testing effort if implemented correctly, and the organisation can reap great benefit from it. The key benefits include higher test coverage, greater reliability, shortened test cycles, quicker releases, consistent regression testing, the ability to do multi-user testing and the support of periodic releases, all resulting in increased level of confidence, stability and consistency in testing.

Test automation can be used in various areas such as functional testing, non-functional testing, test management, performance and volume testing (PVT), scalability testing, soak testing, security and penetration testing, infrastructure testing, technical testing such as unit testing, application programming interface (API) testing, web services testing, build validation, and release and deployment testing. This book is focused on test automation, to help project and product managers, Scrum masters and managers in decision making. This book emphasises test automation through user interface, command line, non GUI, API and many other forms. The chapters also include various other testing concepts and examples that have a significant impact on the overall testing life cycle.

Test automation is not an alternative solution to manual testing, and it requires a strategic view with a tactical approach in order to reap the expected benefits.

This book is divided into two parts. **Part One** addresses the strategic aspects of test automation, and **Part Two** addresses the practical and tactical aspects of automated testing. Test management and defect management tools are briefly covered at various places in this book for completeness and coverage.

Chapter 1 examines what test automation is and why it is required. This chapter addresses the benefits of test automation, advantages, disadvantages and how to manage the stakeholders of test automation. This chapter explores cost-benefit analysis (CBA) and return on investment (ROI) in test automation, and how they help develop test automation policy, strategy and approach.

Chapter 2 describes test automation in different business domains and models such as IT consulting, system integration, multi-vendor, multi-supplier, start up, small/medium sized organisations, and managed test services. This chapter explains the test automation in functional and non-functional testing and the latest trends in the industry such as robotic process automation (RPA) and automation tools based on machine learning (ML) and artificial intelligence (AI).

Chapter 3 describes the factors of tool selection, selection processes and assessment, along with coding and scripting concepts. In order to support the user further, an automation suitability template and a tool evaluation template based on this chapter are included in the Appendix.

Chapter 4 provides an overview of how to build a successful automation test team including team roles, skills, competency matrix, job description and various facts related to hiring the right skills. This chapter is a manager's guide to building a test automation team.

Chapter 5 of this book provides guidance on different automation frameworks, and the advantages and disadvantages of certain frameworks over others. This chapter will help you to understand different test automation frameworks. The generic Test Automation Architecture (gTAA) was developed by the International Software Testing Qualification Board (ISTQB) as part of the Test Automation Engineer (TAE) certification syllabus. This is included in the Appendix G for reference.

Environment is a key factor for the success of any test automation suite and **Chapter 6** describes various environments and impacts of automated testing.

Chapter 7 provides guidelines on identifying the right candidates for test automation. This chapter addresses what should be automated and what should be avoided for automation. This is a key success factor for test automation.

Chapter 8 summarises Part One of the book by addressing how test automation enhances test coverage, for which it is widely used.

Part Two of the book covers various stages of automated testing and how to implement them in an IT organisation.

Chapter 9 addresses how to build a successful career in automated testing, the skills that are required, and the latest trends in test automation.

Tool selection is always difficult as there is a large list of tools available for test automation and automated testing. **Chapter 10** introduces various tools and considerations while choosing a tool. This chapter also lists the leading and widely used test automation tools for reference.

Coding is the centre of any automated testing and it makes automated testing more efficient. **Chapter 11** provides an overview of programming languages widely used for automated testing.

Chapter 12 provides a systematic approach to designing and developing a test automation framework by using a real-life example. This chapter also addresses SUT architecture and test data management.

Chapter 13 addresses how to measure test automation and automated testing with sample reports and graphs to better understand the measurement.

The Appendices are an important part of this book as they list templates, standards and examples from the industry where test automation has been used effectively.

This book is a guide for managers and leaders to introduce and build an effective test automation practice in their organisations and, as a result, help them to achieve better results in test automation and ensure success in software testing.

PART ONE
THE 'WHATS' AND 'WHYS' OF TEST AUTOMATION

1 TEST AUTOMATION: A STRATEGIC VIEW

Software testing is the process used to validate that the software solution or product meets requirements and expectations. Software testing also aims to find defects and demonstrate that the software is fit for purpose. There are numerous testing methodologies, types and techniques available to validate the functional and non-functional requirements.

Test automation is the process of using other software for automating manual testing or manual user actions performed in the application. Test automation helps to automate testing tasks to be performed with the help of tools known as test automation tools. It is a good idea to validate the software with the help of other software where manual testing is either not possible or time-consuming.

Automated testing is validating a software solution using a specialised software tool, and typically involves automating functions as part of the testing process.

This book explores the fundamentals of test automation and automated testing. It is written for senior test automation professionals, project managers, product owners, test managers and Scrum masters to assist with decision making in test automation. In this opening chapter, you will be introduced to the strategic view of test automation.

1.1 INTRODUCTION

Automated testing is the use of special software, separate from the software being tested, to control and execute tests, including the comparison and reporting of actual outcomes to the predicted outcome. The application is known as the application under test (AUT) or system under test (SUT), and the software used for testing is known as the automated testing tool (ATT).

Testing is necessary for all IT systems, and there are numerous instances of IT systems that have gone live without appropriate testing and ended up with defects, causing financial and reputational damage. Testing is a core activity in any IT solution development and is independent of the software development life cycle (SDLC) approaches such as DevOps, Lean, Agile or Waterfall.

IT projects and testing are carried out under three constraints: cost, time and scope. These three factors, commonly called 'the triple constraints', are represented as a triangle (see Figure 1.1). Any changes to the triple constraints create an associated impact on quality, which is measured through testing.

Testing strategies endeavour to create the optimal quality in the solution.

Figure 1.1 Triple constraints

- **Scope.** Scope reduction saves testing time; however, it implies that only a core set of product features are tested.

- **Time.** More time is required for more testing, and this will help to achieve confidence in quality but at an increased cost.

- **Cost.** Cost is directly proportional to time and scope.

This triple constraint is regularly visited from day one of any IT project due to delay of upstream dependencies, and test managers are often under pressure to shorten 'time'. Risk-based testing (RBT) is one of the widely used approaches to optimise testing time and effort, test automation is another. Risk-based testing and test automation can be combined.

Risk-based testing is a good solution for managing quality risks due to the reduced time it takes. Test effort allocation based on the risk of failure is one of the efficient and effective ways to optimise testing. One of the key benefits of the risk-based test approach is ensuring that the maximum value is derived from the planned testing activities, even when the time available to complete test execution is reduced, by the late delivery of the solution, environment or resource availability problems and so on.

Risk-based testing is not about risk and issue management per se – it is about managing quality risks – but builds upon these processes to help identify critical areas that require testing focus. All testing activities need to be prioritised to ensure that the most important tests are completed early so that if time available for testing is reduced, the most critical tests can be completed.

A few methods to manage the triple constraints with minimal impact on quality and testing are listed here:

- Start testing or test preparation early, for example informal testing prior to the planned and scheduled testing.

- Involve the testing team from the beginning as part of the business case development or the design phase.

- Reduce duplicate testing such as common tests in system testing (ST) and user acceptance testing (UAT).

- Join up or merge testing cycles, for example the last cycle of system integration testing (SIT) and the first cycle of UAT.

- Introduce the quality assurance (QA) process as a proactive measure.

- Use testing techniques like exploratory testing to find defects.

- Introduce **test automation**.

Test automation is a software development project that includes most of the phases in the software development life cycle. Test automation is widely enhanced by test automation frameworks.

A test automation framework is a programming framework that includes a comprehensive set of guidelines to produce beneficial results from the test automation activity. An automation framework is either provided by the ATT or, in some cases, a customised tool, which manages test automation in order to produce a better outcome. A test automation framework generally provides a structure to automation tools that fit its purpose. Most test automation tools provide a default framework for automating the SUT; however, the automation tools can be customised for specific requirements, for example an automation framework that schedules various testing tasks as best fit a specific need, generating custom test reports. Test automation frameworks will be explained in detail through the later chapters of this book.

There are numerous ways to make software testing efficient, and test automation tops the list.

Test automation is a key approach in reducing testing effort, but it is not the panacea to all testing activities. Test automation can be introduced at different stages and phases of the test cycle, such as:

- Product development or solution implementation

- Test management

- Functional and regression testing

- Support (post-live) or run

- Test generation

- Test data generation
- Inspection and evaluation of test results
- Compliance

Test automation is widely used across various industries and applications, and it produces great results. Software development approaches such as DevOps, Agile, Waterfall and their different flavours widely use test automation to reduce cost, increase efficiency and accuracy, and speed up regression testing. Test automation is widely used to reduce cost in the long run. Figure 1.2 demonstrates the cost reduction by using test automation over time.

Figure 1.2 Manual testing versus test automation

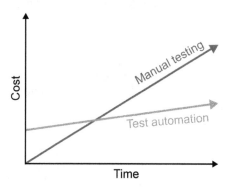

Test automation is a strategic decision that is taken based on sufficient data and analysis. However, many projects and organisations approach test automation as simply a method for cost reduction and end up failing miserably. It is not a short-cut to reduce cost, save time and improve the quality of software testing, and should not be treated as such. Various factors such as SUT, technology and life cycle play a major role before considering automation as part of testing. If used correctly, Test automation is one of the most reliable methods of delivering software testing successfully and securely. There are numerous challenges in the implementation of successful automation, such as tool identification, skill requirements and ways of working.

1.2 MANUAL TESTING AND AUTOMATED TESTING

This section covers how software testing is generally performed. Software testing is performed in two different ways:

- Manual testing
- Automated testing

1.2.1 Manual testing

Manual testing is the process of validating a software solution manually. This often involves a person who mimics the role of an end-user under both normal and unusual conditions. Manual testing also involves technical checks or exploratory testing. Manual testing involves activities such as requirement analysis, risk assessment, identification of scenarios, test case development, test execution, verifying the actual result against the expected result, reporting defects for deviations, updating the test result, and validating the features after defect fixes. The above tasks are performed by a human in manual testing. Test automation tools can be used to enhance manual testing, such as using a defect management tool to report and track the defects. The section below covers some advantages and disadvantages of manual testing.

Advantages:

- Suitable for exploratory testing as this involves creativity and domain knowledge.
- Effective for short-term projects as the testing can be quicker than automated testing, avoiding the time and effort needed to set up a tool and perform the testing.
- Best for single-release projects as there is no need to create, maintain and perform a regression suite for post-live release. Test automation is unlikely to provide sufficient return on investment (ROI) for a single execution. ROI is covered later in this chapter.
- Humans can notice defects that automation overlooks, for example user interface (UI) testing.
- Good for user experience (UX) and usability testing, human intervention is required for these.
- Test execution can start at any time as there is no dependency on tools and infrastructure.

Disadvantages:

- Manual testing is generally time-consuming compared to the time that the same task will take if done in an automated fashion, for example data validation testing of large volumes.
- Some manual testing is tedious to perform: testing a credit card application for many users by filling out the same form time after time, for example.
- Heavy investment in human resources as hiring and training testers requires high levels of effort and investment. Human errors can be avoided using Test automation.
- Debugging complex defects can be difficult and might need multiple runs to collect more information.
- Manual test results are not always accurate due to the higher probability of errors compared to machines.

1.2.2 Automated testing

Automated testing is validating a software solution using a specialised software tool, and this typically involves automating functions as part of the testing process. The most common candidates for Test automation are:

- Test management and defect management
- Unit and unit integration testing
- Functional testing
- Regression testing
- Non-functional testing such as performance and scalability

Advantages:

- Quick and reliable – Test automation makes test execution quick and reliable. It eliminates human error that can arise in manual testing.
- Consistent and comprehensive – The testing will be consistent for any and every data input. The test data set can be increased to 100 per cent coverage. For example in payroll parallel run (PPR), a test automation tool can provide 100 per cent data coverage for all employees in the organisation. Automation tools deliver precise results as testing is accurate, consistent, comprehensive and quick.
- Repeatable – Test automation enables reusability by executing the same tests multiple times with minimal cost. This is useful in extending the test coverage to multiple platforms, environments, operating systems, browsers, and mobile devices. Test automation helps to extend regression testing of different platforms, for example multi-browser, and increases test coverage from development to production. Automated testing can repeat the same tests on different data, platforms and environments with limited resources and fewer overheads.
- Programmable and reusable – This reduces the dependency on manual interactions and helps with automatic scheduling.
- Reduces time for testing – Multiple tests can be executed concurrently, which helps complete testing quickly and reduces the testing timeframe. It also reduces the long-term cost and supports frequent releases of the application by reducing the elapsed time for testing and getting the application to live faster.
- Improves the productivity of human testing – Test automation reduces the manual testing effort. It can achieve 24/7 test execution, more importantly, it can be integrated to run automatically when code is committed. Test automation helps ensure a better use of resources, consistency and repeatability of tests.
- Provides a detailed test log – The test log and execution report are important in testing. Test automation tools can send a detailed report including pass and fail scores with reasons for failure to stakeholders in different locations and time zones.
- Highly recommended for graphical user interface (GUI) and performance testing – Performance testing of large volumes is nearly impossible without test automation. GUI testing on multiple platforms is time-consuming without test automation. Test

automation is highly used for non-GUI, API, service level and API testing and is an integral part of 'pyramid' and 'ice cream cone' models of testing.

- Assured test coverage – The test coverage and accurate measurement of quality is guaranteed in automated testing, which is subject to the quality of the automated test suite used for testing.

- Test automation supports all testing types and phases of testing, including functional testing, non-functional testing and test management – Test automation provides significant payback in the future for tests that are difficult to perform manually and require more time.

Disadvantages:

- Less coverage on negative and exploratory testing – Exploratory testing is an approach to software testing that is often described as simultaneous learning, test design, and execution. This is less formal and not based on structured test cases and steps. Exploratory testing is very effective when performed by a functional subject matter expert (SME) with an awareness of formal testing. Testing tools have limitations in performing exploratory testing.

- Static testing and reviews – Static testing and code reviews are generally performed without executing the code. Testing tools are often limited in performing detailed static testing. However, there are many white-box tools that are effectively used for automated static reviews, for example SonarQube is an open-source platform that is used for continuous inspection of code quality to perform automatic reviews with static analysis of code.

- Tools incur high costs – Automation tools incur several costs including infrastructure, maintenance, development and human resources, even if the automation tool is open source or freeware. Automation suites require continuous maintenance. Tools upgrades or patches also often incur high costs.

- Analysis of the test result is time and cost intensive – Test automation suites can be developed to validate the result manually; however, any failures require detailed analysis to locate the root cause and are time-consuming.

- Requires sound coding skills.

- High maintenance cost – Many functional automation tools are GUI dependent, and maintenance is high if there are any changes to the GUI.

- Not an effective method for short-term products and projects – Automation is generally not suitable for short-term software product development and ROI will be low.

- The application needs to be stable enough to execute automated scripts – A stable application is generally required for most automated testing. Functional automation preparation and execution are largely subject to a stable application.

- Software testing tools often introduce their own defects – The defect can be part of the test automation suite and not a defect in the SUT.

1.3 THE 'WHATS' AND THE 'WHYS' OF TEST AUTOMATION

Test automation is one of the most misused approaches in software engineering. It is very often used as a solution for reducing time and replacement of manual testing. The strengths and weaknesses of Test automation are not often well analysed in many projects. Test automation is neither a solution for all testing problems nor a replacement for skilled manual testers. It is not a solution for reducing testing; instead, it is a long-term investment that pays dividends over a period. Test automation is a software development project that undergoes most of the phases in a software development life cycle. The final solution or product requires continuous maintenance and support. Test automation employs the same software engineering practices that would apply to any software project, and it requires expertise and skills in testing tools and programming languages.

Test automation saves a lot of effort and energy needed for the rigorous testing of the system. It ensures wide and consistent test coverage in the test execution. It helps in achieving 100 per cent validation of large data sets to ensure that no stones are left unturned to provide stakeholder confidence. Test automation can be used to reduce the level of manual testing and manual testers in the long run. Although test automation brings a lot of benefits, the need for manual testing never goes away, but effort could be more focused on exploratory testing, one-off testing (one-time testing) and usability testing.

An example from one of the largest global healthcare portals, with over 500,000 pages and 10 million unique visits a month.

This SUT required the detailed testing of periodic releases, monthly regression testing and data validation testing. The test team consisted of 10 test analysts across functional testing, non-functional testing and test automation. The testing team grew quickly from 2 to 10 members to meet the testing requirements. The application required 100 per cent data validation due to the sensitivity of the information, including data related to public services. The main challenge was the increasing cost and time required to complete the manual regression testing (which took up to 5 days).

A hybrid automation framework was introduced based on a cost-benefit analysis and return on investment. The automation framework helped to ensure 100 per cent data validation and functionality testing on multiple platforms overnight. The good exception handling and reporting modules helped to manage the test execution efficiently and deliver the test execution reports to the stakeholders quickly. A two-member test automation team helped to meet the break-even point within six months and 100 per cent test coverage.

1.3.1 Why test automation?

Test automation provides the following benefits to an organisation and its customers and stakeholders:

- Increased test coverage and better testing quality – Test automation can extend test coverage by performing testing of a large volume of data. Test automation can also provide consistency and accuracy in test execution by eliminating human error.

- Improved efficiency and effectiveness – Test automation can repeat the same test or the same test with different inputs with less cost and effort, bringing increased efficiency in overall testing.

- Better reliability and accuracy – Test automation is more reliable than manual testing as it reduces human error from test execution.

- Cost-saving and time reduction – For example, automated regression testing helps to run the regression tests quickly, more often, within different environments, unattended and with large data sets. This will reduce the cost and save testing effort.

- Test automation enhances testing as it can run unattended, reducing dependency on time, place and resources.

Figure 1.3 shows the benefits of test automation and why it is required.

Figure 1.3 Why test automation?

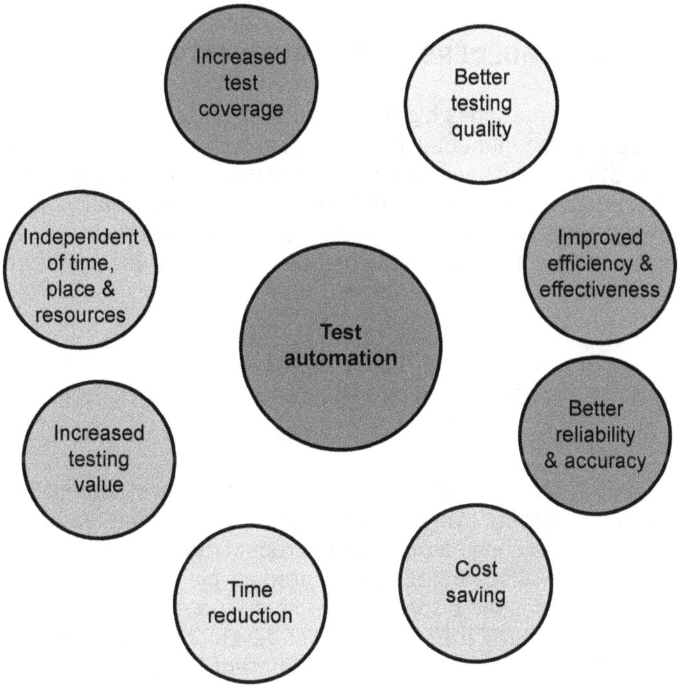

1.3.2 Testing types and test automation tools

Testing can be divided predominantly into two types: functional testing and non-functional testing. There are a large set of open-source, custom-developed and vendor-provided tools available to support functional and non-functional testing. There is also a third category of testing tools for test management activities such as test planning, test design, defect management, and reporting.

- **Functional test automation.** Functional test automation tools automate the tests concerning application functionality or how applications function.

- **Non-functional test automation.** Non-functional test automation tools enhance non-functional requirement testing, that is, the performance, usability, accessibility, reliability, security and infrastructure and so on of a software solution. Most of these non-functional requirements are difficult to test manually. For example, testing a website for thousands of concurrent users is not possible without an appropriate testing tool.

- **Test management.** Test management is everything the test team does to manage the software testing process, such as test planning, test preparation, test case development, test data preparation, test execution, test matrix, test reports and defect management. Test management tools help test teams to manage the testing process, test life cycle and reporting.

Now we'll explore another key factor for the success of test automation, stakeholder management.

1.4 MANAGING STAKEHOLDERS IN TEST AUTOMATION

The success of test automation is highly dependent on support from the stakeholders, as automation requires a continuous and long-term commitment to funding, resourcing and sponsorship. Stakeholders vary in different organisations and engagements. In most product-based companies, the immediate stakeholders for automation are internal such as a product owner, product lead, development lead or product manager. There are also generally external stakeholders such as customers, vendors and system integrators.

> A stakeholder is any individual, group or organisation that can affect, be affected by or perceive itself to be affected by a programme, product or SUT.

Although testing is a standard phase in the software development life cycle, it is managed differently in product versus multi-supplier (MS), multi-vendor (MV), system integrator (SI) and IT consulting companies. In product development organisations, test automation budgets are allocated as part of product planning. Here testing tools are mostly procured or established prior to the product or solution development and they are already in place for many matured product development organisations as a part of the Test Centre of Excellence. However, it is different in multi-supplier, multi-vendor, system integrator, and IT consulting environments, where test automation scope, requirements, tools etc. are identified during the bespoke solution planning, design, and development phases. Implementing test automation is often influenced by the cost, reusability, post-live support, and internal/external stakeholder priorities. Therefore, managing these factors becomes the primary focus for managers and test leads prior to commencement of test approach. IT business models are explained in Chapter 2.

The matrix in Figure 1.4 explains how stakeholders are to be involved based on their stake in the solution along with their interest in the solution.

Figure 1.4 Stakeholder involvement matrix

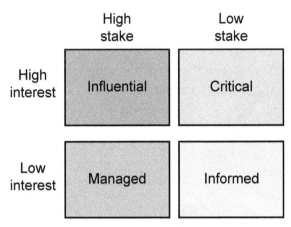

Below is a set of factors to consider in managing both the internal and external stakeholders in the test automation process.

- **Stakeholder identification.** Identify all the stakeholders at the beginning. Stakeholders fall into different categories such as business, IT and technology, legal, project management, suppliers and trusted partners, product or project team, infrastructure and environment team, or tool vendors. It is important to work well with all these stakeholders for the success of test automation. For example, tool vendors and open source tool communities are an integral part of the success of test automation as they often provide timely support or professional services when the automation team faces technical challenges.

> Note that some of these external stakeholders may be professional services that incur additional costs.

The role of the stakeholders and their support to test automation vary based on their stake in the automation. For example, a business stakeholder may be the right person to provide test data for automated testing and an IT stakeholder may provide the right support for a test environment for automated testing, such as providing the infrastructure for automation, granting access to the servers, designing and building environments, and backing up the data. All stakeholders must agree and be clear on their role and responsibilities, and clarify their priorities.

Figure 1.5 represents the stakeholders involved in test automation. The key stakeholders have a direct impact and are usually the primary beneficiaries of the solution; hence they are the centre of the stakeholder landscape. Other stakeholder support is required to deliver a successful solution and test automation.

Figure 1.5 Stakeholders

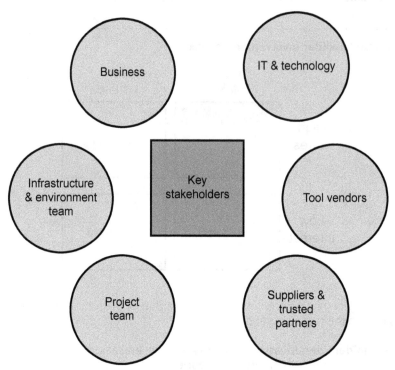

- **Goals.** All projects have many priorities; however, stakeholders have their individual objectives and goals. It is vital to get an understanding of the key objectives of the stakeholders to plan how best to manage them. A good understanding of the key priority of the prime sponsor helps to define the automation approach accordingly.

  The rationale for stakeholder goals and business priorities regarding test automation generally fall into one of the following:

  - Complete the product or project implementation within the timeframe.
  - Being first to market to get an edge on the competition (reduce the testing window).
  - Cost reduction or controlling the cost.
  - Long-term return on automation investment.
  - Team reduction (e.g. reducing the number of manual testers).
  - Reuse of the existing automation tool licence and avoid procuring new tools.
  - Build automation capability internally and avoid external support.
  - Consistent and stable regression testing to increase confidence in releases.

A leading global provider of agricultural science and technology, in particular seeds and crop protection products, hired a global UAT test manager for a compensation management system that was implemented in over 90 countries. The UAT manager was on-boarded in August. Following a series of meetings with various IT and business leaders, the test manager listed the business, human resources (HR), data and IT objectives. After walking through the top three priorities in the meeting with the chief technology officer (CTO) of this European multinational agro company, it was understood that the primary priority of the key stakeholder (CTO) was to ensure that the tool would be available for production in December (timeframe) to adhere with the business needs for the next year. Any delay would impact the annual performance appraisal and the tool would not be useful for another 12 months, with the manual system continuing to fulfil the need for the current calendar year.

The testing scope and objectives were immediately reviewed and reprioritised to meet the key stakeholder's goal.

- **Communication.** Communication channels, frequency and methods are to be defined and agreed upon with the stakeholders at the start of the product or project implementation. Stakeholder communication should be value-based and differentiated based on their stake, that is, avoid the one-for-all approach to stakeholders. For example, when providing periodic reports to the stakeholders, senior stakeholders may be interested to see the 'traffic light' or red, amber and green (RAG) report with highlights, and product or project stakeholders may be more interested in details of the test automation progress.

- **Stakeholder engagement.** It is important to engage the stakeholders throughout the test automation as they will have a special interest in how it is shaping up. Keeping them in the dark and ignoring their concerns will reduce or stop their buy-in. There should be no element of surprise for any stakeholder on a given day.

- **Agree on the outcome.** Agree with the stakeholders on what is expected to be achieved, what are the benefits and return on investment, and provide evidence that the test automation is on track to achieving them. Any change in plan or delay in achieving benefits should be communicated on time.

- **Avoid overselling.** Avoid overselling and providing inflated benefits of test automation to the stakeholder to procure budget. For example, avoid overselling or inflating the benefits of the test automation tools to justify the tool costs.

- **Consult and consult again.** The benefits of test automation, particularly in the early stages, may be unclear to its stakeholders. Regular stakeholder consultation is essential to ensure that requirements are delivered to agreement.

1.5 TEST AUTOMATION POLICY, STRATEGY AND PLAN

Test automation policy, strategy, plan and approach ensure test automation is designed and implemented as agreed. This is a key success factor in test automation.

A policy is a document described at the organisation level and provides insight for the activities at the highest level. The purpose of test policy and test automation policy is to ensure that adequate direction is provided to all parts of the organisation, and associated third parties, in relation to testing and test automation. The policy also outlines the standards required when performing software testing and test automation.

Test automation policy, test strategy and test plan can be separate, stand-alone documents or a combined document, subject to the size of the organisation and practice.

The document hierarchy of stand-alone documents is represented in Figure 1.6.

Figure 1.6 Document hierarchy

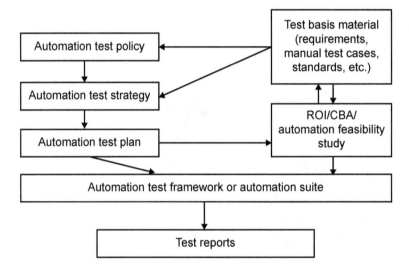

1.5.1 Test automation policy

A policy is a statement of intent and is generally adopted by a governance body within an organisation. Test automation policy provides guidance to ensure the industry standards and the best practices in test automation are followed to achieve the most efficient and effective outcome. Every test organisation and organisation that undertakes software testing (tech product and project-focused organisations) should have a test automation policy, which should adopt some of the key principles below:

- Test automation activities are carried out with proper governance and control.
- Test automation complies with test process maturity models.
- Test automation follows coding standards and guidelines.
- ROI should be calculated for all test automation activities.

- Automation feasibility analysis should be conducted for all test automation (this is explained later in this chapter).

- Cost-benefit analysis (CBA) on automated testing should be performed before a test automation approach is concluded (CBA is also explained later in this chapter).

- Automated testing should be deployed as much as possible where the benefit of cost saving can be realised.

- Test automation should be used to maximise productivity and cost efficiency.

Refer to Appendix E for a sample test policy template.

1.5.2 Automation test strategy

The test strategy is a document describing the approach to the testing of the artefacts and is the top-level plan generated and used by the testing team to direct the test effort. Test automation strategy describes 'what' the test automation is for the solution and 'why' test automation is required. It provides a summary of the test automation and responsibilities specific to the life cycle of the automation. It may be high level, dependent upon the availability of test automation requirements, and could further be elaborated in the following document, that is, the test automation plan. Test automation objectives, scope and resources will be detailed within the strategy document.

The purpose of a test automation strategy is to provide a rational deduction from high-level organisational objectives to actual test activities.

The main content of the automation test strategy is as follows:

- Purpose of the document
- Solution information and background
- The scope of test automation and why it is required
- High-level automation approach
- Test automation tool approach
  - Tool selection approach
  - Proof of concept (PoC) approach
  - Proof of technology (PoT) approach
- Phases of automation
- Test automation framework (high level)
- Environment details (high level)
- Roles and responsibilities
- Stakeholders that will be responsible, accountable, consulted and informed (RACI)
- Risks, assumptions, issues and dependencies (RAID)

1.5.3 Automation test plan and approach

The test plan comprises separate documents for each test phase or a master test plan with all the phases and subprocesses in line with the test strategy, including all the information necessary to plan and control the test effort for the solution development phase. It describes the approach to the testing of the artefacts and is the top-level plan generated and used by managers or leads to drive the test effort.

The automation test plan describes the scope, approach, resources and schedule of intended test automation activities (for example functional test automation plan or performance test automation plan). The test automation strategy explains the 'whats' and 'whys' of test automation, and the test automation plan describes the 'hows' and 'whens' of automation, mainly about how and when the solution will be automated and executed. It also identifies, among other test items, features to be automated, automation tasks, the task owner, test environment, test automation techniques, coding standards, and entry and exit criteria to be used. It is a document of the test planning process. This is a living document of how test automation is to be implemented in a product or project, updated often to ensure that the plan and approach are up to date.

The test automation plan consists of an approach that is the implementation of the automated tests for a specific solution.

The main content of the automation test plan is as follows:

- Purpose of the document
- Solution information
- High-level scope of automation
- How and when automation is performed
- Detailed automation approach
- Details of the selected tool
- Phases of automation implementation
  - Initial
  - Preparation
  - Final
- Test automation framework (detailed)
  - File structure
  - Features
  - Libraries
  - Reporting
- Guidelines and standards
- Test environments and data

- Stakeholders that will be RACI

- RAID

The sequence of activities and deliverables in the test automation approach are represented in Figure 1.7.

Figure 1.7 Automation test approach

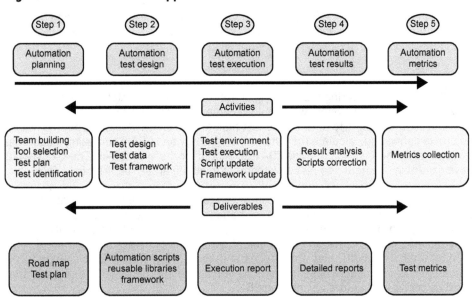

Refer to Appendix E for a sample automation test plan and template.

1.6 CBA AND ROI FOR TEST AUTOMATION

Test automation has various impacts on many areas of software testing. It can save costs and resources while also resolving many testing-related problems. But there are many horror stories where test automation has failed and large investments have been shelved, causing customer dissatisfaction and ultimately calling off the automated testing. Understanding cost versus benefits and computing ROI for test automation will help the leadership team in deciding to invest in automation tools for testing. CBA is one way to do the ROI calculation.

1.6.1 Cost-benefit analysis

CBA is a relatively simple and widely used technique for deciding whether to make an investment in test automation. As its name suggests, simply add up the value of the benefits and subtract the costs associated with it. The automation costs are described in more detail later in this chapter. Costs can either be one-off or ongoing. Benefits are most often received over time. The standard justification for choosing test automation over manual testing is that the benefits exceed the costs over the life cycle of the product or project. Doing an actual analysis of the benefits and costs can encourage better decision making and ensure that resources are allocated effectively to support test automation.

A cost-benefit analysis is the process to measure the benefits of a decision or an investment minus the costs associated with that decision or investment. CBA is widely used for business and investment decision making.

The CBA should start during the planning phase (before implementation) for any project. As the requirements develop, we should be able to determine testing methodologies or techniques and start to develop the test plan and strategy. This early outline will be helpful to define and calculate the automation strategy and CBA. The testing leadership is responsible for ensuring that the CBA is executed for the automation. The lead should have expertise in automated test development in all the areas required for a CBA.

There are many factors to be considered when planning for a software test automation and analysis of cost and benefits. Automation changes the complexion of testing from design through implementation, test execution and test result analysis. The development, maintenance and execution of automated tests are quite different from manual tests. The skills, test approaches and, moreover, testing itself change when automation is practised. These impacts have positive and negative components that must be considered before performing CBA.

The major factors that should be considered for CBA for test automation are as follows:

- Project specifications or product requirements – The requirements are major factors to be considered for CBA. For example, the products designed for large user bases or multiple releases can impact the test automation approach and related benefits. A product based on a set of complex requirements may be difficult to automate, and it may incur long-term maintenance cost.

- Functionality to be tested – The functional and non-functional features. Some functionalities may not require any testing. For example, a commercial off-the-shelf (COTS) product may require minimum testing compared to a custom developed product.

- Iteration or release of the software solution and testing – A product that requires multiple iterations in the future can be a suitable candidate for automated regression testing and provide a high return in the long run.

- Software testing and testing methodologies – Costs and benefits of these are influenced by the software development life cycle. A DevOps-based product development requires frequent testing in multiple environments, and manual testing may not be feasible or incur a high cost.

- Payback for the automation investment – This is a key factor for CBA calculation. The payback is estimated based on many assumptions, and it needs to be revisited if any of the corresponding assumptions are changed.

- Schedule variance and performance impacts in testing when automation is introduced.

- Future iterations of the SUT – The number of future iterations with the life cycle of the SUT and amount of testing.

- Tool patches, updates and so on – Tool patches, patching windows, frequency and professional support for patching are major factors to be considered.

- Resources available (e.g. hardware, tools, people) – The resources required for any kind of testing need to be considered for CBA. It is important that the right resources be available for estimating CBA.

- Reusability of previous test structures – This can reduce the cost while increasing the benefits. However, previous structures can incur additional maintenance costs.

- Test strategy and plan – The test strategy and plan will provide an indication of the level and type of testing required for the solution in its life cycle.

- Automated scripts maintenance – This is a largely ignored area in test automation. The automated test scripts require changes subject to SUT changes, enhancements or new releases of the testing tool and so on.

CBA framework

The first step in performing a CBA is to define what you are testing and what you plan to automate. This will invariably come from the product or project plan and test plan as well as previous automation effort.

Figure 1.8 explains the CBA framework. Benefits are estimated based on tangible and intangible factors, and they can be impacted by the risks and assumptions. The testing can be performed manually or by automation or a mix of both. The work breakdown structure is generally used to further clarify the potential structures. Detailed data must be collected for each potential structure to estimate costs and benefits. Some possible sources of data are product knowledge, past projects, domain knowledge, current automation costs and online whitepapers.

The steps for a high-level CBA framework, which are explained in detail in the following sections, are:

1. Identify potential test structures

2. Ascertain assumptions and risks

3. Estimate benefits: tangible and intangible benefits

4. Estimate costs

5. Convert the costs and benefits into comparable values

6. Compare the alternatives and decision making

Figure 1.8 Cost-benefit analysis framework

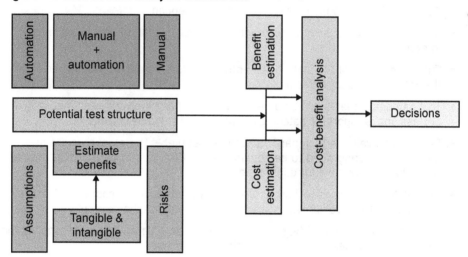

Identify potential structures The standard justification for choosing one level of testing over another is that the benefits exceed the costs over the life cycle of the project. Doing an analysis of the benefits and costs of using various testing structures can encourage better decision making. There are three different structures that should be considered when performing a CBA, and the one chosen should be the most cost-effective within the context of budgetary and timeline considerations. The structures are:

1. Automated testing (only automation)
2. Manual testing + automated testing
3. Manual testing (only manual)

When comparing structures, it should be ensured that the same level of test coverage has been considered for all structures. At this point, some structures can be rejected because they are not feasible.

Ascertain assumptions and risks As we are analysing these structures, we will be making a lot of assumptions, and there will be risks. It is important to identify the assumptions and validate them on the basis of fact or prior experience. For example we may assume that there will be one operating system and three service or security packs to this version of the SUT. Based on our track records and data, we will decide whether feasibility and maintainability will be an issue. This also provides a platform to explain why some structures were dropped in the analysis. If we did consider and eliminate some structures early on because of a conclusion that would not be feasible, the assumptions behind that conclusion should be clearly explained and justified.

Estimate benefits The benefits of any automation can be tangible or intangible. They are often mixed and sometimes difficult to measure. It is important to assess tangible and intangible benefits and assign a value or weightage to them. This can be positive such as cost reduction and quick testing. Likely tangible and intangible benefits are listed below.

Tangible benefits:

- Human involvement – Automated tests can reduce human involvement during testing, saving time, rather than manually running the same tests.
- Code coverage – Code coverage can be used to estimate test effectiveness. Automated tests can be incredibly effective, giving more coverage and visibility into the functionality of the SUT, for example testing an SUT for a large set of data.
- Test execution time – Automated tests can reduce the test execution time as computers are often quicker than humans.
- After-hours testing of systems – Automated testing helps to run the tests 24/7.
- There is a reduction in manual testing time for the current iteration and for future iterations.
- Automation of testing that would otherwise be impossible, for example load testing.

Intangible benefits:

Intangible benefits are difficult to assess but are a key factor for CBA and are generally intertwined with tangible benefits.

- Hands-off testing – Although the reduction in the cost of people is easily measurable, any additional value of saving in computer usage is difficult to quantify and is an intangible benefit.
- Good reputation of test organisation – This often increases the productivity of testing as the reputed test organisations have a high acceptance for their approach, established standards, rich in-house expertise and reusable test automation solutions (TASs) in order to assist the test automation.
- People – Not all members of the test team will want to change. Some turnover in personnel often accompanies test automation, even when some testers prefer to continue to test manually.
- Change in quality – The quality can be better after automation as automated testing can improve test coverage and consistency.
- Number of cycles of test execution – Automation often allows faster confirmation of product builds and provides the additional opportunity for the organisation to run the same test more frequently. This may not be feasible for manual testing.
- Payback – There can be unexpected payback from automation over a long period such as quick releases and reduction in testing cost. An immediate payback may be seen for some automation, for example build validation tests, but usually, some payback comes much later after the investment.
- Increase in reliability – Automated testing can be more reliable as it can avoid human errors due to the boredom of running the same test again and again.

People are different, and the quality of manual test execution is dependent on individual capabilities.

- Satisfied customer – The test evidence from automated testing is more consistent and reliable. This can increase internal and external customer satisfaction.

Estimate costs   Financial costs associated with automated testing can be generally described as either fixed or variable costs. Fixed costs of automation are expenditures for equipment, tools, training and so on. The variable cost increases or decreases based upon the number of tests that are developed or the number of times the tests are run. The following factors will be helpful in addressing the costs:

- Automated tests may generate mountains of results that can need much more staff involvement for analysis, thus costing more to run than manual tests.
- Time for automation planning, development and execution.
- Personnel cost is a key factor for automation costs. People with prior experience in automated testing are required for successful test automation, and experienced personnel increase the overall cost of testing.
- Hardware resource cost such as infrastructure required for tool installation, execution, configuration management and test execution machines.
- Testware such as software licences or software support such as professional support from the community and tool vendors.
- Automation environment maintenance such as test automation script backup.
- Tool introduction, training and ramp-up such as training from the vendor, training on programming languages and cost of certification.
- Test execution costs such as how often automated tests can be run, resources required for execution, result analysis and licence cost for execution.

In some cases, data may not be available to provide an adequate cost estimate. In that situation, the best alternative is to use the judgement and previous experience of the team members to estimate costs.

Convert the costs and benefits into comparable values   After the costs and benefits for each structure alternative have been estimated, you will have a mixture of tangible (monetary) values and subjective level of benefit. It is generally harder to assign the benefits a financial amount. If there is no realistic way to relate the value of the intangible benefits to the tangible ones, they cannot be considered significant for the cost analysis. In this case, they can be used as a decision maker if the comparison of structures does not show that one structure is a clear winner. The effort should be converted to level of effort (LOE) and a monetary values assigned to it.

Compare the alternatives   At this point, the analysis can proceed by comparing the net benefit (benefits minus costs) for each structure to determine which structure is best for the current project. The options with the greatest net benefit should be selected.

Refer to Appendix E for a cost-benefit analysis template.

1.6.2 Return on investment calculation

In general, automated testing involves higher up-front costs than manual testing. Performing ROI analysis for test automation will help to determine up front what types of automation are ideal for the project, what tools will be required and what level of skill will be required for the testing. Not only does ROI serve as a justification for effort, but it is also a necessary piece of the planning process for the project. Projects that do not perform ROI calculations up front do not fully understand the costs of their automation effort, what types of automation they could be doing versus what they are doing, and what strategies to follow to maximise their return. The ROI is perhaps the single most effective way to garner the interest of decision makers. If you have compelling numbers backed by solid facts, and if you can show that every pound spent on test automation will return two pounds down the road, for example, then you are practically guaranteed funding.

Return on investment is a ratio between net income and investment. ROI is widely used to choose investment options. ROI measures the amount of return on a particular investment.

The ROI of test automation over releases is presented in Figure 1.9.

Figure 1.9 ROI versus releases

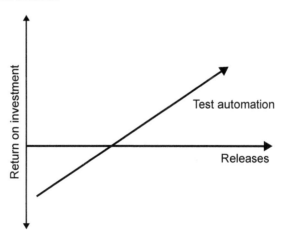

When it comes to automating the test process, the costs are tangible. But the net present value also includes many intangible factors (as we have already discussed in CBA). The findings in CBA can be used for an ROI calculation. The best approach is to determine the benefits with as much precision as possible and then compare them to the cost of automating the test effort.

A few different formulas to calculate ROI are:

1. ROI = Return/Investment
2. ROI = What you get out/What you put in
3. ROI = (Benefit – Cost)/Cost
4. ROI for automation = Benefits from automation over manual/Cost of automation over manual
5. ROI for automation = Cost of automation – Cost of manual testing

ROI calculation from the insurance industry based on formula 5

Business purpose: verify the functionality and validation of the base system for entering life insurance policies.

Cost of automation = Cost of hardware + Cost of software + Cost of test preparation, that is Test script development + (Test scripts maintenance x number of times scripts are executed) + Cost of test execution (Test script execution x number of times scripts are executed) + Cost of test result analysis + Defect reporting and tracking.

Cost of automation = £XXX.XX

Cost of manual testing = Cost of test preparation, that is Test case development + (Test case maintenance x number of times tests are executed) + Cost of test case execution, that is Test case execution x number of times tests are executed + Test result analysis + Cost of defect reporting and tracking.

Cost of manual testing = £YYY.YY

ROI for automation = Cost of manual testing – Cost of automation (£YYY.YY – £XXX.XX).

Note: A positive value shows there is positive ROI for test automation.

Refer to Appendix E for a ROI template.

1.7 AUTOMATION FEASIBILITY

The organisation decides to progress the wide adoption of automated testing once the CBA and ROI are deemed reasonable. To achieve the best test efficiency, automation will be considered during the solution planning. The above analysis on automated testing should be carried out and presented to the stakeholders for approval on the investment. However, it is recommended that each team looks into its solution for automation suitability as not all solutions are suited for automation. This will help the organisation and the individual solution to use automated testing to reduce the cost of testing and improve efficiency. Automation suitability study at an individual product level will help the team to assess whether testing activities such as test case design, test data setup and test execution are suitable for automation.

The following factors are considered for automation feasibility or suitability analysis:

- Solution platform – A solution or product development for multiple platforms such as browsers, operating systems or environments will be highly suitable for automated functional and regression testing. However, the automation suite should be independent of platforms.

- Frequency with which the SUT code changes and is released – It is a good indicator for automation feasibility if the products require frequent testing. However, the automated pack requires frequent maintenance due to changes in the SUT.

- Tools availability and compliance – Availability of the SUT compliance tool is a key factor to be considered for automation feasibility. Lack of an automation tool will increase the effort and cost to develop an automation suite for testing.

- PoC or PoT – Proof of concept and proof of technology will help to understand the suitability of the SUT for automation (refer to Chapter 3 for further details).

- Resource availability – Availability of hardware, software and human resources is another key factor for automation feasibility.

- Benefit realisation – CBA supports benefit realisation for test automation.

- Duration of test execution – Automation may help to reduce the long manual execution cycle.

- Execution frequency – Frequently executed tests score high weightage in automation feasibility analysis.

- Technical complexity – Technically complex SUT and testing tools tend to increase the effort for automation.

- Live solution duration – The life cycle of the application is another factor for feasibility analysis. A SUT with long life is a good candidate for test automation and qualifies for feasibility analysis.

- Test data availability – Availability of test data for execution and less manual intervention to feed the data for testing will increase the weightage for feasibility.

- Dedicated environment for test automation – This provides a test environment available for automated testing and reduces the dependency on other teams.

- Requirement stability – Stable requirements will reduce the effort for automation.

- Detailed manual test steps – Detailed manual test steps reduce the effort for test automation preparation.

- Business as usual (BAU) team availability for handover – A dedicated post-live support team is a major factor for test automation. Lack of a support team will make the automated suite obsolete.

Refer to Appendix E for automation suitability templates.

1.8 OVERSELLING AUTOMATION

One of the key problems in test automation is overselling to senior leadership. Over-projected stories from automation tool vendors and inflated success stories from other organisations presented to senior leadership can be detrimental later in the cycle.

'Many organisations have not been able to get the level of Return on Investment from automation initiatives they would have wished for' according to the *World Quality Report 2019–20* (Capgemini, 2019).

In a rush to automate the solution and secure the investment, test automation is often oversold to the stakeholders. Overselling can have a negative impact on the success of implementing test automation across the organisation. This raises doubts in the stakeholders' minds, and they may no longer trust the decision-making ability of the test managers. The common pitches used to oversell test automation are:

- Once test automation is achieved, testers will not be needed.
- One hundred per cent of automation can be achieved.
- Test automation should be implemented to meet the tight deadlines.
- If your automated suite is passing at 100 per cent, there are no defects in the application.
- High and quick ROI.
- It requires a single click to execute test cases through automation.
- Test automation will help to find more defects.
- Automated testing is always better than manual testing.

1.9 SUMMARY

Test automation is a subset of software testing, and it increases test coverage and supports a large volume of data testing. However, there are numerous challenges and complexities in test automation and, without overcoming them, it will be difficult to build and manage a successful automation function.

In this important opening chapter, we addressed what test automation is and why it is important in testing and product or project management. We also covered the decision factors in test automation and the advantages of automated testing over manual testing. This chapter also covered the importance of ROI and CBA in test automation and how to calculate them. Additionally, we addressed stakeholder management in test automation. Next, we will further uncover how test automation works in different IT business models, domains, testing types and testing stages.

2 DOMAIN-FOCUSED TEST AUTOMATION

Test automation is the validation of a software solution using other software. There are lots of failures in test automation, the main reason being the limited experience of test managers across various domains and sectors. Test automation itself is an IT project, and no two IT projects are the same. Many test automation projects end in failure as the team tries to replicate the automation success from a previous project to the current project without considering the various factors that are salient to the current engagement.

This chapter explains how test automation is applied across different domains, sectors and development approaches such as Agile, Waterfall, DevOps, continuous integration (CI), Lean, Kanban and so on. It aims to assist test managers, Scrum masters and automation leaders in understanding the challenges and factors to be assessed before commencing automation planning. In the following sections, test automation is further explained in different SDLC approaches, business models, testing stages or subprocesses, testing types and the latest trends in test automation.

Table 2.1 provides a synopsis of this chapter.

Table 2.1 Test automation and domains

Domains	Area covered
SDLC or development approaches	Agile, Waterfall, DevOps, CI.
Business models	IT consulting, project, product, MS/MV, SI, managed testing services (MTS), start-ups, SMEs.
Testing types	Functional and non-functional testing.
Testing subprocess (stages)	System testing, integration testing, regression testing.
Latest trends	Artificial intelligence (AI), machine learning (ML), robotic process automation (RPA).

2.1 SOFTWARE DEVELOPMENT APPROACHES

This section covers test automation in Waterfall and Agile software development approaches.

Software solutions are developed through different approaches, and most of them follow a flavour of either traditional Waterfall or Agile approaches. Test automation itself is an IT solution, and hence it should follow an approach. In most cases, test automation is part of a wider IT programme, and it follows the programme's software development approach. However, many organisations consider test automation itself as an IT project, and this allows them to follow a development approach of their choice.

> The software development process is a methodology that describes the activities involved in defining, building and implementing an IT system.

The Waterfall testing approach follows the same structure as the Waterfall software development approach wherein there is a sequence of stages in which the output of each stage becomes the input for the next stage. Testing is performed step-by-step in line with the implementation steps, subject to the completion of a phase or stage.

Figure 2.1 shows the different steps/stages of the Waterfall approach.

Figure 2.1 Waterfall approach

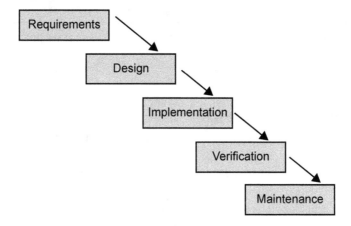

Agile testing is based on the Agile development approach, where testing is not a different phase and is performed alongside development phases that include requirements analysis, design, coding, test case development and automation.

Figure 2.2 shows a pictorial representation of the Agile development and testing approach.

The Waterfall and Agile testing approaches are fundamentally different. Table 2.2 compares Waterfall and Agile testing.

Figure 2.2 Agile approach

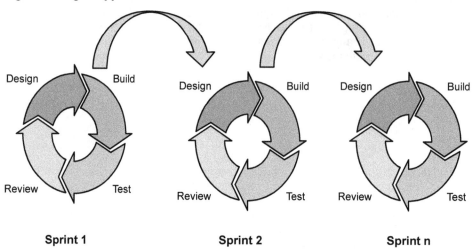

| Design | Build | Design | Build | Design | Build |

Sprint 1   Sprint 2   Sprint n

Table 2.2 Comparison between Waterfall and Agile testing

Waterfall testing	Agile testing
Testing is separate and in distinct phases, such as system testing, system integration testing, performance, volume testing and UAT.	Testing is not a separate phase or stage in Agile testing.
Formal testing is performed only after the completion of development or a stage.	Testing is performed alongside the development.
The development and testing teams work separately.	The development and testing teams work together.
Testers may or may not be involved in the requirement definition.	Testers are involved in requirement refinement and estimation.
Test automation is subject to project need.	Test automation is an essential part of Agile.
Acceptance testing is a separate stage and performed at the end.	Acceptance testing is performed at the end of each iteration.
Big teams involved.	Small team and more collaboration.
Regression testing is carried out only after defect fixes or releases.	Regression testing is carried out frequently.
Long-term planning.	Short-term planning.
Distinct testing phases.	Overlapped testing phases.
Defined and structured implementation.	Flexible implementation.
Majority of the testing is left to towards the end of development.	Frequent testing throughout development.

In the real world, many large projects are implemented using the hybrid approach, that is, Waterfall and Agile.

A large government department designed and implemented an intranet portal based on a Microsoft (MS) SharePoint-based cloud platform. The implementation team followed different development methodologies as appropriate. The SharePoint tool configuration, implementation, functional testing and test automation followed Agile-based development methodology. Infrastructure and non-functional testing used a more suitable Waterfall-based testing approach for the solution.

Functional testing was based on two weeks' Sprint (Agile approach), including fully automated regression testing with Microsoft Team Foundation Server (TFS) and Azure DevOps toolsets. However, performance testing was based on the Waterfall approach fully agreed and signed off requirements using the Micro Focus LoadRunner tool.

The following sections cover test automation in Waterfall and Agile along with another popular approach, DevOps.

2.1.1 Test automation in Waterfall

The Waterfall approach consists of step-by-step phases, where each phase depends on the deliverables of the previous one. This approach is widely followed in large projects. Waterfall requires clear structure and documentation up front. It is divided into distinct stages or steps with quality gates to finish the current stage and start the next one. The stages are relatively standard and often follow a sequence: agree on the requirements and scope, design, implement, test, deploy and post-live maintenance (see Figure 2.3). There is less flexibility with this approach; what is agreed at the beginning must be seen through. However, this approach reduces uncertainty and increases the ability to work on a clear scope and requirements. Changes are often addressed by the change control (CC) process, and it reduces the overall cost.

Test automation in Waterfall projects is easy to implement as the requirements, test cases and a stable application is available for automation (Figure 2.4).

Test automation and Waterfall characteristics:

- Application is available for a proof of concept for test automation – This reduces uncertainty and supports the tool selection and tool compliance for test automation.

- Quick to estimate and calculate ROI as the requirements are stable and agreed.

- Faster test automation as the application is available – An existing and stable application enhances test automation. A stable GUI is essential for GUI-based functional test automation, and it reduces the rework of test scripts.

- Scope and approach for test automation is clear – This allows test managers to plan and estimate test automation and reduces the uncertainty, as the requirements are agreed upon prior to test automation commencing.

Figure 2.3 Waterfall testing

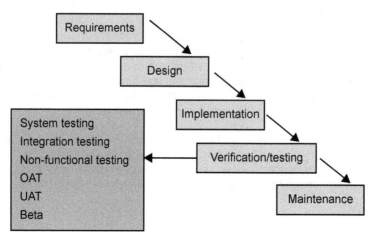

- Cost of test automation is under control – The stable application, agreed requirements and scope help test automation deliver in line with the agreed plan. This reduces rework and cost and provides additional control in test automation implementation.

Limitations of the waterfall approach in test automation:

- Requirement changes are difficult to manage – The Waterfall development approach and Waterfall-based test automation provide limited opportunity to absorb changes. The requirements are agreed upon prior to test automation commencing, and any changes to the requirements are often addressed through a CC process. This limits the opportunity to respond to the changes and feedback.

- Lack of flexibility – Waterfall is known as a structured process and lacks flexibility. Waterfall-based test automation starts after agreeing on the requirements, framework, architecture and schedule, and there is no or limited room for flexibility.

- Not usable for ongoing application development – Waterfall-based automation is not suitable for ongoing continuous integration and continuous delivery (CI/CD) and DevOps-based development as the test automation suite is not delivered frequently and not built to address ongoing development.

- Test automation commences only after application or functionality development – This reduces the ability to use the automation suite during the product development stage.

- No real-time use of ongoing testing, as automated testing commences only after the development is complete.

Figure 2.4 Test automation in Waterfall

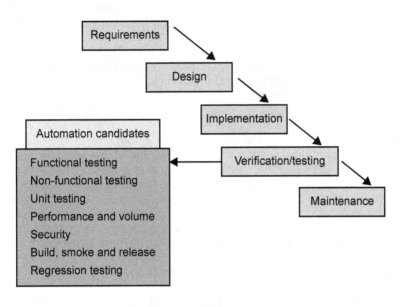

There is a leading scientific and technology software company, headquartered in the United States, with representation in Europe and Asia, providing software for chemical, material and bioscience research for the pharmaceutical, biotechnology, consumer-packaged goods, aerospace, energy and chemical industries. The products are highly scientific in nature, and requirements are agreed upon prior to implementation. The requirements are stable and require minimal changes during implementation.

The products are frequently regression tested in multiple cycles during development stages and post-release. The automated test suite is used for regression and release testing of migration, porting, web, stand-alone, client–server, product testing and cross-platform testing.

The automated regression suite is developed using the Waterfall development approach due to the nature of the applications and the stable requirements for testing and test automation. Due to the nature of the product, platform and release, the automation suite provides almost 100 per cent coverage on regression testing with minimal maintenance.

2.1.2 Test automation in Agile approach

The Agile development approach was developed as a response to growing frustrations with Waterfall and other highly structured, inflexible approaches. This approach was designed to accommodate changes and produce software quicker. Agile emphasises working as one team, customer collaboration throughout the development approach and focuses on creating working software rather than documentation. In the Agile approach, project teams develop the solution and a list of deliverables in short Sprints or iterations of 2–3 weeks in general as illustrated in Figure 2.5, based on the business and technical priorities. During development, teams work towards the goal of delivering working software and review the progress, status and outcome of the deliverable at the end of the Sprint. This is a feedback-driven process and is well accepted by the business community. Agile is collaboration-focused and client satisfaction is the priority in the approach. However, Agile projects sometimes lack control when managing costs, requirements, scope and discipline, resulting in unsatisfied customers.

Agile test automation is an approach to using test automation in Agile development with the purpose of making testing more effective and in line with the Agile way of development. It reduces the dependency on manual testing, uses test automation to speed up the application development, reduces the manual testing effort and uses the team better. This approach accommodates changes. Agile automation aims to automate everything feasible and to replace manual testing rather than using automation for additional benefits. Many Agile test teams tend to automate the features of previous Sprints and use the automation test suite for continuous regression testing.

Test automation and Agile development characteristics:

- Test automation is parallel to product development – Many Agile teams tend to spend the first part of the iteration on requirement clarification, manual testing and so on, and the second part of the same iteration for automated testing. This allows the team to build good knowledge of the features, test them manually and ensure they are stable prior to test automation. This enables faster test automation and frequent execution.

- Agile automation supports ongoing product development by performing automated build validation, smoke testing and regression testing.

- Quick ROI and customer satisfaction as the result of test automation is immediate, within the same Sprint or iteration.

- There is high reusability and support of ongoing releases – The automated test suite is used time after time, mostly daily regression testing and build validation.

- There is high coverage on requirements and testing – Agile automation aims to automate everything feasible and provides high test coverage.

- Long-term costs are lowered by reducing the manual effort for frequent test execution.

- Solution and testing requirements are met quickly as the test automation is parallel to product implementation.

Figure 2.6 shows Agile testing and test automation characteristics. Testing is the core of Agile development and plays a significant role in the inception, construction and transition phases. Test automation enhances the development and testing in the Agile approach.

Figure 2.5 Agile testing

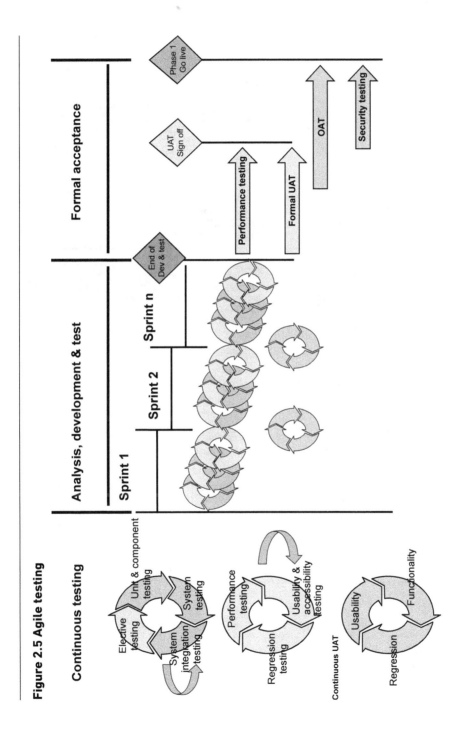

Figure 2.6 Test automation in Agile at an enterprise level

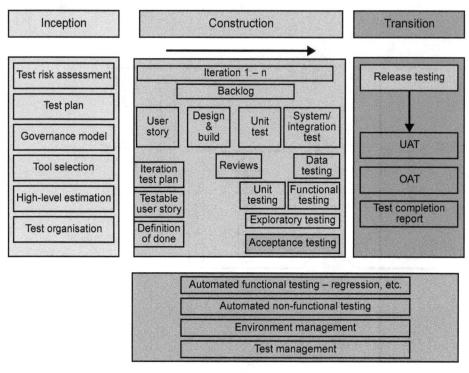

Limitations of the Agile approach in test automation:

- Focus is on getting test automation done in time when it is needed, and not how effective or efficient it is.

- Test automation tends to focus more on functional testing, and non-functional testing is often ignored.

- High maintenance, as automated test scripts require continuous maintenance in line with the Agile delivery.

- The application continually evolves, iteration after iteration, and therefore due to lack of a stable application, testing is difficult to automate.

A leading travel and transport provider implemented identity and access management (IAM) for their external customers who use multiple commercial applications. Customers used to register and maintain separate accounts to access each of the applications and services they are interested in. The IAM solution allows customers to use a single identity with associated personal details and preferences, providing access to a range of applications.

The implementation took a phased approach to the application of IAM and was delivered using a Disciplined Agile methodology. The product backlog contained requirements that were delivered through a number of user stories. The product backlog is a living set of requirements accountable to the product owner.

The minimum viable/valuable product (MVP) was defined during the inception and the solution was delivered over a few deployments. The product backlog and user stories that form the test basis were captured in the Microsoft Azure DevOps tool. Lightweight test plans were created for each iteration based on the user stories and acceptance criteria selected in that iteration backlog. Test cases were developed to validate the acceptance criteria of the user stories, enabling requirement traceability. Testing was based on the IAM architecture and Selenium was used for test automation. Regression testing was 100 per cent automated.

2.1.3 Test automation in DevOps

DevOps is a set of practices that combine software development (Dev) and IT operations (Ops) together, thus the name 'DevOps', or 'development operations'. It aims to increase the rate of solution development with frequent releases utilising many practices from the Agile development approach and Lean principles.

DevOps is a set of tools and key practices that increase the ability to deliver applications at high velocity over traditional software development and infrastructure management processes such as Waterfall and Agile.

Figure 2.7 explains the close association of development, continuous releases and continuous testing in DevOps. Test automation plays a key role in the success of DevOps and is inevitable.

In DevOps style, programmers and test automation engineers work together, supporting the test team in developing an automation framework to perform DevOps testing. The development team contributes to test script development as there is just a thin line of separation between developers and automation engineers. The automated scripts and codes are supported by CI/CD tools, generating builds automatically, deploying and testing them.

The scope, plan, pipeline and production release are defined in advance in order to plan and implement DevOps testing. There are many frameworks available for DevOps implementation and testing, such as Microsoft Azure DevOps Services, Amazon Web Services (AWS) Developer Tools and Google Cloud.

Figure 2.8 explains the CI/CD DevOps workflow and test automation in DevOps.

Figure 2.7 DevOps

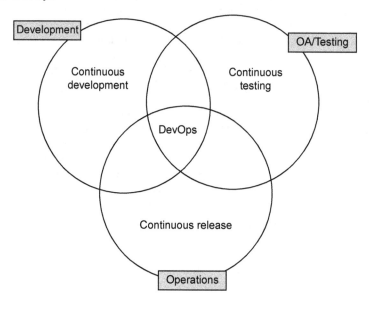

Figure 2.8 DevOps CI/CD workflow

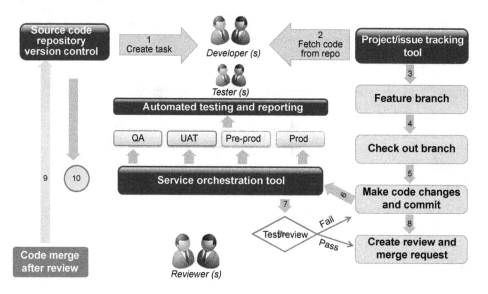

Test automation and DevOps characteristics:

- Code and configuration changes are continuously tested in DevOps. Test automation is the key to achieve this goal.

- Feedback loops are fast and detailed. The new features are continuously tested by the automation suite, and this provides immediate feedback on quality. Test automation shortens the lead time between fixes.

- Automated non-functional testing suites are part of the pipeline and run early in the cycle.

- Automated end-to-end functional testing integrates well with testing frameworks.

- Close association of the development and automated testing team leads to effective cross-functional teamwork and reduced resource dependencies.

- Test automation is a key focus area in DevOps, and it is essential to achieve DevOps goals. This lowers the failure rate of new releases to live and minimises disruption of releases.

- Speedy delivery, high flexibility and instant scalability of additional features and functionality as test automation enhances the testing efficiency and shortens the testing timeframe.

The list above covers a few common goals for both Agile and DevOps, although they are not achieved as often in Agile. However, DevOps gives more emphasis to ensure these goals are met.

Figure 2.9 explains the close association of developers, testers and automation SMEs in the DevOps model. They often interchange their roles in DevOps. A fully working test automation is a key success factor for DevOps.

Figure 2.9 DevOps test automation

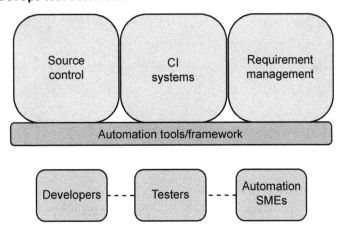

Limitations of DevOps in test automation:

- Continuous and frequent release to production is a core objective of DevOps. However, production rollback could be cumbersome and messy in a few instances.

- Continuous updates to the production system are not always feasible and are difficult for legacy code.

- High scripting and programming skills are essential for the success of DevOps. It requires heavy investment in tools and hardware.

- Mindset for cross-functional collaboration is essential for the success of DevOps, and any human intervention or conflict slows down delivery.

- DevOps is challenging in projects and in products with highly integrated environments, and continuous maintenance of test automation is required.

- Not suitable for highly regulated industries or secure sectors as any release to production requires auditing by the regulators, prior approval and so on.

A large European 'digital retail' organisation decided to build a set of tools that would enable the company to increase retail sales.

The programme features were:

- An online shopping site with minimal IT and marketing investment

- The development and test environments hosted within the cloud using a software as a service (SaaS) or infrastructure as a service (IaaS)

- The solution development followed a DevOps model

The following toolsets were considered to support the DevOps:

- Deployment – Jenkins, Puppet

- Continuous integration – Jenkins, Selenium

- Source control – GIT

- Development environment – Vagrant, Puppet

- Configuration management – Puppet, Ansible

- Monitoring – Azure Control Center, App Insights

- Issue/defect tracking – Visual Studio Team Services (VSTS) or Azure DevOps

- Planning – VSTS or Azure DevOps, Microsoft Project

- Collaboration – Confluence wiki, SharePoint

2.1.4 Continuous integration

CI is one of the original twelve practices of the extreme programming (XP) development process. XP is a software development methodology that was developed by Kent Beck. XP was one of the first Agile methods, as well as being the one that is predominantly used before Scrum takes the lead. CI is a software development approach where the work is integrated frequently. It is also a DevOps best practice and allows developers to frequently merge code changes into a central repository. The code merge or check-in is verified by an automated build and test, that ensures that problems are detected earlier.

CI is a modern software development practice in which incremental code changes are made frequently and reliably.

The developers integrate their code at least daily, though often more frequently. The CIs are validated by automated tests with the intention of finding integration errors as quickly as possible. One of the key benefits of CI is that the automated build and test reduces integration problems and helps the team to develop and deploy the software more rapidly. The automated CI and CD improves code quality and thereby security.

A CI server such as 'CruiseControl' is useful for code deployment, as it builds a custom continuous build process, monitors the code changes, and alerts errors, even though CI does not require any tooling in order to be deployed. There are many on-premise, cloud, and hybrid tooling solutions available for CI that perform tests on every change and monitor the repository for changes. There are many licensed, procured, free, and open source CI automation servers such as Jenkins, TeamCity, GitLab CI, Bamboo, Buddy and so on, which are widely used for facilitating CI and CD to build, test, and deploy software.

Automation is a key activity in CI. The automated test suite validates the code changes and integration to the main and shared code repository. Performing automated tests on every branch helps identify issues early. Automated testing is key for the success of CI as the test automation suites provide immediate feedback to the developers on their product. The automation suites are generally split in two; the first set provides quick feedback (e.g. unit tests), while the second set provides more detailed test feedback. The first set of quick tests always helps the developers reduce the waiting time for the test result.

One key aspect of test automation in CI that the team needs to ensure is that the feature that gets developed has corresponding automated tests, and tests are upgraded in line with new releases, and regression issues. The developers need to be involved as early as possible in the definition of the user stories to create good tests. The test-driven development (TDD) is widely practiced in CI as the tests are created before writing the code that will fulfil the test.

The generic CI steps are as follows:

- **Step 1.** Create tests (e.g. TDD) for the key features of the solution and the codebase.
- **Step 2.** Run the tests automatically (e.g. using a CI server) on every code merge to the main repository.
- **Step 3.** Developers integrate code changes frequently or as soon as possible.
- **Step 4.** Fix errors and the CI build as soon as it is broken.
- **Step 5.** Write new tests to enhance the test base for new stories.

CI was initially created for Agile development. CI works well with the 'Test Pyramid' developed by Mike Cohn and implements test automation at various levels.

- **Unit tests.** Unit tests are performed at micro levels, which verify the functionality of individual methods or functions. Unit tests can be automated largely.

- **Integration tests.** Integration tests (unit integration) ensure that the components work as expected together. This involves the integration of units, classes, and components with other services. Integration tests can be automated largely and are highly usable for CI.

- **Acceptance tests.** Acceptance tests focus on business features and scenarios.

- **UI tests.** UI tests ensure that the application functions as expected from an end user perspective.

CI environment

A CI/CD environment is similar to that of a production environment in which ongoing automation, monitoring, testing, deployment, and delivery takes place. CI aims to reduce code integration and merging issues. The objective of the CI environment is to avoid integration challenges and keep code ready for the production environment with support from test automation. In a CI/CD and DevOps/CD environment using an Agile approach, development and operations collaborate closely together. The code changes are moved through a CI/CD pipeline designed to continuously integrate and deploy new product fixes, updates, and features. In the past, CI has used a separate tool for code build, monitoring, deployment, etc. However, many of the latest CI solutions on the cloud provide a fully integrated CI/CD solution.

The key benefits of CI are:

- It has a great focus on test automation to validate and ensure the product is not broken whenever new codes are integrated into the main repository. This saves time by identifying and addressing conflicts early on. This also helps develop a good automated regression suite for the software.

- CI helps to refine the code base through continuous integration and testing. CI also helps achieve high code coverage.

- CI reduces the scope of late integration defects and helps to identify conflicts early in the product development life cycle.

- The role of manual testing is reduced in CI. This helps QA engineers to allocate more time for debugging and tool support.

- CI reduces risk to software deployments as the releases are proved time after time on the CI server prior to production deployment. CI ensures the product is ready to be released at any time, so it is on demand and there are nearly zero downtime deployments.

- CI allows software products to reach the market much faster. The traditional phased software delivery lifecycle takes weeks or even months for a fully developed code to be in the market. CI helps the team work together, automate the build, carry out automated testing and risk-free deployment and ensure speedy environment provisioning.

- CI improves quality as the automated tools help to achieve quick regression testing. This helps the team find more time and effort for other testing activities, such as exploratory testing, usability testing, performance testing and security testing.

- Metrics generated from automated testing and CI (e.g. code coverage, code complexity and defect trends) help the team to improve overall quality.

CD involves packaging the software for deployment in a production-like environment. The primary objective of CD is to ensure that the software is always ready to go to production. The organisations that adopt CD can release numerous releases into production frequently, subject to passing all automated tests, and provide good test coverage. CD also enables roll back changes to take place quickly if something goes wrong.

2.2 TEST AUTOMATION AND BUSINESS MODELS

Software testing and test automation is planned and managed differently in different IT and business models such as IT consulting, SIs, MS/MV and MTS. In this section, we will address test automation in different business models.

2.2.1 IT consulting

Information technology, consulting firms and consultants advise on the planning, design and implementation of information technology systems for their clients. IT consulting focuses on advising organisations how best to use information technology in achieving their business objectives. IT consultants need strong interpersonal, technical, business and communication skills to deal with clients. IT consulting services are advisory services that help clients assess different technology strategies. Most consulting firms specialise in the advisory role, although many of them are supported by their system integration units to implement solutions. The well-known IT consulting firms, such as IBM, Accenture, Deloitte and Capgemini, have strong IT consulting arms along with system implementation expertise.

Testing and test automation are core parts of IT consulting. Some of the IT consulting firms offer testing as a service (TaaS) or testing consulting services from the advisory role to implementation. This section addresses how consulting firms support the client test automation process. Test consulting firms have a pool of test automation skills, and most of them offer this as a service. Experienced automation consultants from these firms bring expertise in various areas of automation such as feasibility studies, calculating ROI, managing test automation, implementing automation, independent testing, managing other test automation service providers and filling the skills gap.

Test automation and IT consulting characteristics:

- Identifying the right TAS – IT consulting companies have tremendous experience and expertise in various tools, tool implementation, automation strategies and frameworks generally used in a similar environment. This helps to avoid reinventing an automation solution from scratch and reduces the risk of failures in test automation.

- Tool procurement – Most IT consulting firms are partnered with tool vendors. This helps them to procure the tool quickly and cheaply. The tool procurement can be managed by the IT consulting companies with an additional marginal cost.

45

- Identifying and vetting the right external test automation partner – This involves identifying similar automation solution partners, assessing them and prioritising on behalf of the customer.

- Sourcing expertise in automated testing – This involves providing professional testing services' resources from within the IT consulting company or externally.

- Advising on various testing services such as on-premises, cloud, TaaS and hybrid – This is mainly from IT consulting companies' vast experience in test automation with different clients on different platforms.

- Building the automation framework – IT consulting companies have considerable expertise in this area and often reuse the expertise, components and framework from previous projects.

- Performing a test automation maturity assessment – IT and testing consulting companies have vast expertise in this area and have standard templates to perform test automation maturity assessments. They often have certified resources to assess test automation maturity.

- Implementing test automation processes and procedures – IT consulting companies generally have a pool of resources they can augment to deliver the test automation need for the customers.

- Advising on infrastructure for test automation – Many IT consulting companies perform an advisory and solution provider role for infrastructure for test automation and TASs based on IaaS, platform as a service (PaaS) and SaaS.

Limitations of test automation in IT consulting:

- Limited knowledge of client's business operating model and infrastructure landscape – IT consulting companies and resources have sometimes limited knowledge of the client's business and needs. It is difficult to build knowledge of the customer's business quickly. They often tend to provide a standard TAS to the customer rather than a custom-made solution.

- The stakes on automation success and ROIs are often highly inflated – IT and test consulting companies tend to move to a 'sale' mode (sell the service) rather than focusing on a successful TAS. The solution is intertwined with the upselling objective.

- Test automation consultants are expensive – IT or testing consulting is often expensive as there is a premium charged on their reputation and expertise.

2.2.2 System integrator

A system integrator is generally an IT company that specialises in bringing together component subsystems into a whole with a group of similar providers with different skills. They deliver testing and test automation as part of the solution. System integrators primarily operate on two financial models: namely, fixed time and fixed price (FT/FP) or time and materials (T&M). In fixed price, they often agree upon the scope of the work, the timeframe to deliver and the cost with the customer up front, with changes managed through change control. The time and materials model is generally used when the scope is not clear, and it is difficult to size the work.

Test automation works slightly different in T&M and FT/FP. In an FT/FP, the SI has a good amount of knowledge of the scope of testing, test automation, tools to be used, amount of work, test automation framework, timeframe and ROI. They mostly support both the solution development and post-go-live releases and continue maintaining the automated solution for the BAU phase. The fixed-price project generally includes the delivery of the first release of the system, and the test automation could even be omitted to reduce the cost. However, the test automation framework could be designed with the view to stay long term to support BAU activities if the SI is responsible for post-live maintenance of the solution. Most of the T&M models are approached with a different view as the test automation tools and tasks are not defined up front. Test SMEs use the T&M model to mitigate the unknowns of test automation.

Test automation and system integrator characteristics:

- System integrators have good knowledge of test automation. They have huge experience in delivering multiple systems across a variety of clients. SIs have expertise in end-to-end test automation, including functional and non-functional test automation.

- SIs have an existing partnership with major tool vendors and providers (such as Gold or Platinum partnership). They have the ability to procure tools on behalf of the customer at a reduced cost.

- They have a large pool of skill sets available within the organisation and prior experience in producing quick TASs and estimating ROIs.

- SIs have wide experience in automation solutions, tools and infrastructure. They have experience in defining test automation responsibilities.

- SIs have experience in the selection of the appropriate tool and automation approach and expertise in estimating the cost and benefit of tool usage. SIs also have the skills in training people to use the automation tools.

Limitations of test automation in system integrators:

- Lack of innovation and mainly template-driven (cookie-cutter model). SIs try to implement the same solution in all places to reduce the cost.

- Partner with a close network of vendors and tend to sell their products with a margin.

2.2.3 Multi-supplier and multi-vendor

An MS and MV model is where the client uses different products and solutions across a multitude of suppliers and companies on the same platform in order to develop a bespoke solution for a specific business need. The overall solution is managed and maintained by the customer, and they procure sub-solutions from different vendors. MS/MV requires more effort as there are a lot of different providers, solutions, platforms and infrastructure components on the same landscape. It is challenging to coordinate activities as some of these suppliers can be large consulting companies, IT service providers and product vendors. A single provider creates a dependency on one supplier and leads to high cost and opportunity loss; therefore, MS/MV avoids using solutions from a single vendor. A few major challenges in an MS/MV environment are:

- The customer's IT organisation should be strong and mature enough to manage multiple suppliers.
- Unhealthy competition between solution providers.
- Risk of small vendor business closure or insolvency.

Test automation principles in MS/MV are very similar to those of the system integrator, enabling the customer to procure the optimal automation solution for their testing.

Test automation and MS/MV characteristics:

- Tool procurement and maintenance is the customer's responsibility. However, the customer has a wide choice of automation tools and automation offerings from the suppliers.
- Suppliers can procure tools with less cost on behalf of the customer as they have a long-term relationship with tool vendors.
- Access to different service models, such as TaaS and MTS, is available for the customers through the MS/MV. This provides easily scalable and multiple test automation options. A pay-as-you-go TAS is also available for the customer through the MS/MV.
- Test resource augmentation through the suppliers is possible in MS/MV and this helps by requiring less time in setting up a test automation suite.
- Suppliers are experienced in similar solutions for other customers and reusability of the solutions is highly possible.
- Technical know-how and experience with previously used templates and solutions is available through MS/MV.

Limitations of test automation in MS/MV models:

- Large MS/MV environments are cumbersome, and generally lack support from other suppliers to automate the testing.
- Competition between suppliers is not always good.
- Suppliers are at times reluctant to support automation if they do not have any stake or financial benefits.

2.2.4 Product-based test automation

Software products are applications, solutions or systems delivered to the customer with documentation that describes how to use them. A few examples of software products are:

- Application software, such as Microsoft Office and Symantec Endpoint Protection
- Operating systems, such as Windows, macOS and Linux

- Internet browsers, such as Chrome, Safari and Edge
- Mobile applications, such as Instagram, WhatsApp and Skype
- Large enterprise applications, such as enterprise resource planning (ERP), AWS and Microsoft Azure
- Games for all platforms, such as Roblox and Minecraft

The products are mainly created for multiple customers, and some products are customised for specific needs. Most of the software products are managed through the licence agreement, whether paid or unpaid. Product testing is different from the models discussed above in this chapter, as it involves some specific testing such as installation testing, patch release testing, localisation (L10N) testing and so on. Test automation has a major role in the success of the product, and ROI is high in product-based automated testing. Test automation adds huge value to product development and support.

Test automation is widely used in the product development environment. The products are generally tested on different platforms, and manual testing is cumbersome. The automated scripts can be used for different platform combinations, versions and patches. Security patches and the latest releases should be verified on different platforms. Regression testing is not possible without automated scripts.

Test automation and product characteristics:

- Test automation helps in achieving test coverage quickly on different platforms and environments.
- Test automation enhances repetitive, release and regression testing of regular upgrades, patches and hotfixes.
- Test automation enables a large user volume, and the stress testing required in product testing.
- Automated scripts can be used for many versions or releases in product testing.
- Installation testing can be quickly validated by test automation.

Limitations of test automation in product development:

- Continuous test suite maintenance requires product testing, and this involves additional effort and cost.
- High maintenance of the test automation suite is required due to additional requirements in upgrades, patch releases, multiple platforms and environments.
- It is less effective over a period as bugs may become immune to the automation suite.

2.2.5 Managed testing services

MTS cover all activities in a test project, including the entire testing process. Most of the IT offshore testing providers offer different flavours of MTS. Test automation is a key part of MTS. MTS and TaaS are used to represent the same concept, as both models grant on-demand access, ready to use testing services, high-quality consultants and proven solutions.

Figure 2.10 lists the common services offered by MTS models.

_____

Figure 2.10 MTS pillars

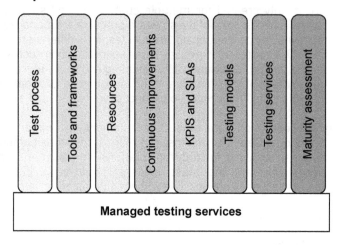

Note: SLA: service level agreement

'Managed Testing Service is an end-to-end, fully customised service in which we take responsibility for test activities at either enterprise or programme level. A collaborative approach to sourcing testing and QA services, MTS is the solution for clients who want to transform their testing function and achieve high-quality systems cost-effectively, without the expense of large overheads and day-to-day responsibilities' (Capgemini, 2014).

Test automation and MTS characteristics:

• Test automation is fully managed by the MTS provider and test automation can be predefined and pre-agreed in MTS.

• Test automation is managed and performed onsite, offsite and offshore, helping to save significant cost.

• High efficiencies, particularly in test automation and performance testing, and reduced deployment time.

• MTS offers fully managed test service and resource availability.

• Frees up in-house staff, enabling a focus on core business challenges and objectives.

- MTS offers training as a service.
- A shift from capital expenditure to usage based on operating expense.
- Total flexibility to use as needed (subject to the price), reducing fixed testing overheads.
- Access to increased test automation capacity.
- MTS can provide advanced test environments and access to the latest tools, devices and platforms for test automation.
- Measurable quality improvements in the MTS model.

Limitations of test automation in MTS:

- Constant liaison with the business for decision making, and a high dependency on in-house resources, for example data for test automation
- Loss of in-house skills and expertise in test automation
- Hidden costs such as additional tool features and test automation dashboards
- Numerous change requests for enhancing test automation scope and additional features

2.2.6 Start-ups, small and medium enterprises

Start-ups are generally new companies founded by one or a few aspirational individuals to create a unique product or service. They are in the initial stages of business and so mostly face the challenge of getting off the ground while simultaneously trying to attract investment. There is a high expectation for start-ups to get their products or services in the market with limited resources. Therefore, in most start-up organisations the employees tend to stretch their roles to various areas beyond their titles. A similar establishment is small/medium scale businesses, which are enterprises or corporations which generally operate with a low initial investment or have a small number of employees.

Start-ups and small/medium sized businesses often try to optimise resources. They have similar characteristics, such as limited hiring experience, lack of an established HR department, established company reputation, limited resources, new IT department, new tooling team and so on. However, in general the whole team is passionate about the business and want it to succeed.

Most of the small or medium business houses are driven by market demands and so are the first to market. They often lack IT tools, processes, and documentation, for example a small sized company may use a spreadsheet or a free open source tool for defect management compared to a licensed and procured popular test management tool. They often focus on immediate needs, such as quickly fixing the defect and moving onto the next one rather than gathering test metrics. Reusability is a key ethos in start-up and small/medium sized organisations, where test automation plays a key role. They tend to reuse the existing solutions available in the market with minimum rework. Software is often developed and tested with agility. They often avoid detailed planning, documentation, and developing an automation framework, and instead start to automate and deliver the working solution on day one. Testing is important for start-ups that focus on 'technology first' where the business depends on having an application or solution that delivers what it is selling.

Documentation is a core part of any IT project. The level of documentation varies across projects, programmes, products, companies, and industries. In start-ups and small/ medium sized organisations, the documentation is often a few lines in an email or a wiki page rather than a detailed 'change controlled' process. The documentation is important in testing and test automation. However, it should be realistic and fit for purpose. Start-ups and small/medium sized organisations tend to focus on a 'what is fit for purpose' or 'immediate objectives' approach during their documentation process.

The characteristics of test automation in start-ups and small/medium organisations are as follows:

- Agile-based development, testing, and test automation is often practiced. The focus is on immediate goals.

- Software testing and test automation is vital at every stage of development. CI, Lean, Kanban and DevOps are often practiced.

- Products are developed based on MVP, which is generally followed to get a sense of user engagement.

- Reusability of testing strategies and test automation is crucial as it reduces cost and helps in delivering the solution quickly.

- Testing and test automation are important as they have very little space for error.

- Free and open-source software are very popular due to low cost and ease of setting up.

Limitations of test automation in start-ups and small/medium organisations are as follows:

- Limited budget and time to test the solution on every possible permutation, combinations, configuration of devices and environments.

- Less focus on quality in the process and documentation.

- Testing and test automations are often side-lined due to focussing on delivering a working solution.

2.3 TEST AUTOMATION IN FUNCTIONAL AND NON-FUNCTIONAL TESTING

Functional testing validates the expected features of a solution, and non-functional testing validates the scalability and sustainability of the solution. Both are success factors for any solution and supported by automated testing.

2.3.1 Functional testing

Functional testing validates the solution against functional requirements and specifications. The purpose of functional testing is to verify that the solution works as expected in line with requirements. The key types of functional testing are:

- Unit testing
- Unit integration testing
- Smoke testing
- System testing
- Integration testing
- Regression testing
- Exploratory testing
- User acceptance testing

The functional requirements for a solution are generally described through the products shown in Table 2.3.

Table 2.3 Functional requirements specifications

Product	Description
User stories and epics or business requirements document	Statement of high-level functional requirements
Requirements specification	Statement of functional requirements
Interface definition documents	Detailed interface specification(s) for each interface
Wireframes	Page designs
User experience design	Look and feel page designs based on wireframes
Product backlog	Comprehensive list of requirements used to drive the detailed design and development activity

Figure 2.11 shows the key areas of test automation in functional testing.

2.3.2 Non-functional testing

Non-functional testing is validating the solution for its non-functional requirements, the way a system operates, scalability, sustainability and other software quality characteristics. Functional and non-functional tests, testing types and stages are often used interchangeably because of the overlap in scope between requirements. For example, security testing is often used for security accreditation validation, penetration testing and vulnerability scanning, and performance testing is a broad term that includes

Figure 2.11 Functional test automation types

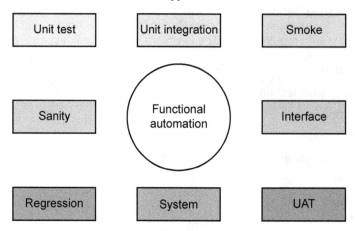

many specific requirements such as load, soak, volume, response time, capacity, reliability and scalability. A functional test, for example, is to check a report is working as expected, displaying the correct data as per the calculations, and a non-functional test is to check that the report loads within a defined time, for example 5 seconds. The key types of non-functional testing are:

- Usability testing
- Performance testing
- Load/volume testing
- Scalability testing
- Availability testing
- Supportability testing
- Maintainability testing
- Security testing
- Regulatory and legal testing, for example accessibility testing
- Compliance, standards and policies testing
- Access testing, for example single sign on (SSO)
- Compatibility testing
- L10N and internationalisation (I18N) testing

The non-functional requirements for a solution are generally described through the products shown in Table 2.4.

Table 2.4 Non-functional requirements specifications

Product	Description
User stories and epics or business requirements document.	Statement of non-functional requirements.
System architecture or high-level design.	Architecture views of the solution, including non-functional requirements.
Interface definition documents.	Detailed interface specification(s) for each interface.
Component design architecture or low-level design.	Detailed architecture for each component.

Figure 2.12 shows the key areas of test automation in non-functional testing.

Figure 2.12 Non-functional test automation types

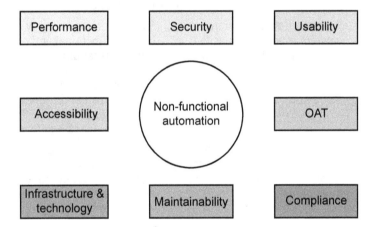

2.3.3 Testing stages

The following testing stages and types are performed generally as part of the solution development. The functional and non-functional tests, testing types and testing stages are often used interchangeably because of the overlap in scope between requirements.

- Unit testing/component testing (UT/CT) and component integration testing (CIT)
- System testing (ST)
- Factory acceptance testing (FAT)
- System integration testing (SIT)
- User acceptance testing (UAT)

- Performance and volume testing (PVT)
- Operational acceptance testing (OAT)

Table 2.5 describes the test stages, types and respective test automation.

Table 2.5 Testing stages and automation

Test stage	Key objectives	Test automation
Unit testing or component testing and component integration testing	To prove that each developed software component (or package configuration) functions according to its technical specification and that deployed unit components interoperate as expected.	Unit testing and unit integration testing are largely automated. There are various tools and packages available to automate unit testing. A unit test can be 100% automated and provide high returns. A few widely used unit testing tools are JUnit, xUnit.net, NUnit, TestNG, PHPUnit, Symfony Lime, Test Unit and RSpec.
System testing	To test that the solution or system meets the specified functional and non-functional requirements for different types of valid or invalid conditions.	System testing is highly automated for both functional and non-functional requirements. There are many tools and frameworks available to automate both functional and non-functional requirements such as UFT, Selenium, Tosca and Eggplant.
Factory acceptance testing	To ensure that the solution is fit for deployment into the customer's environment for full-release testing.	FAT is subject to the business domain and how different each installation is. Optimally, all automated tests can be executed to validate the installation (site or factory).
Systems integration testing	To ensure that the solution integrates and functions in end-to-end internal and external systems with no data loss or corruption.	This is similar to system testing and highly automated for both functional and non-functional requirements. There are many tools available to automate both functional and non-functional requirements.

(Continued)

Table 2.5 (Continued)

Test stage	Key objectives	Test automation
User acceptance testing	To verify that the solution is fit for purpose from a business perspective, i.e. business requirements and user acceptance criteria have been met. This is usually performed by the end-users or business users, or their representatives.	UAT generally follows manual testing. However, payment, security, data migration and parallel process testing are usually automated as it is not feasible to test them manually. In Agile, a large part of UAT tests are automated as acceptance test-driven development (ATDD) and behaviour-driven development (BDD) are derived from the epics or user stories.
Performance and volume testing	To verify that the solution meets the specified performance requirements when placed under different levels of load, e.g. under normal, peak and extreme workloads. Load variations may include the number of users, the volume of transactions and the volume of data.	Most performance tests require functional or performance testing tools to run effectively. Some commonly used performance testing tools are: • LoadRunner • JMeter • Silk Performer
Operational acceptance testing	To provide operations and support staff with confidence that the system operates correctly in the production environment and can be managed by them. Tests typically focus on support readiness, testing issue resolution through the service desk, and operational support aspects like backup, archiving, monitoring or alerting, recovery point objective (RPO), recovery time objective (RTO), maintenance, disaster recovery, scalability, reliability and availability.	OAT is performed using a mix of manual and automated testing. Backup, archiving, monitoring, alerting etc. are widely automated using various tools. However, disaster recovery, failover etc. are generally performed manually.

2.3.4 Testing types

According to the International Software Testing Qualifications Board (ISTQB) definition, testing types are a means of clearly defining the objective of a certain level for a programme or project. The tester focuses on a particular test objective during test case execution. Table 2.6 describes the different types of testing and the role of test automation.

Table 2.6 Testing types and automation

Test type	Objectives	Test automation
Static testing and static code review	Analysis and review of test base material (e.g. functional requirements, non-functional requirements, high-level design (HLD), low-level design (LLD), full system architecture, detailed design architecture) to verify that the building block specifications of the solution are unambiguous and provide sufficient information.	This is largely manual testing; however, test management and version management tools bring discipline to this activity. An automated review comment tracker ensures comments are addressed quickly and reduces manual effort. Tools such as Microsoft SharePoint, Microsoft Azure DevOps or Micro Focus Application Lifecycle Management can be used for review comments tracking. Static code reviews are supported by tools like Git or Bamboo.
Functional tests	To test that the delivered functionality within the application meets the specified requirements.	Functional tests are largely automated subject to the ROI, benefits and qualification of the right test automation candidates. Some common tools used for automated functional testing are: • UFT • Tosca • Selenium • Eggplant • Appium
Regression tests	To test that the application has not been negatively impacted by the introduction of new functionality, changes or fixes, i.e. tests to verify that what was working before still works.	Same as functional testing, regression tests are largely automated subject to the ROI, benefits and qualification of the right test automation candidates. The same tools for functional test automation are used for automating regression testing.

(Continued)

Table 2.6 (Continued)

Test type	Objectives	Test automation
Business cycle tests	To emulate the activities performed by the application over time periods such as daily, weekly or monthly and ensure that the application performs as specified end-to-end.	Test automation adds high value to business cycle tests such as periodic reports, payroll and parallel processing. There are specific tools available for business cycle test automation alongside functional automation tools. Microsoft Excel with macros is very useful in this testing.
Data migration testing (DMT)	DMT is also known as extract, transform, load (ETL) testing. This is to verify that the data routines specified are operating correctly and that extracted data can be loaded onto the new platform with no loss of integrity of data. This tests that the data load and migration processes function properly.	Data migration is certainly a candidate for 100% automation. Manual data testing is prone to error and not feasible for large migrations. This is largely automated testing. The widely used data migration testing tools are: QuerySurge, RightData, BiG EVAL, iCEDQ and Informatica Data Validation.
Load testing	Testing of the system's performance for designated transactions or business functions under different workload conditions and across different time periods. The following should be reflected in the performance testing: • Response times (interactive work) for end-users, defined by averages or percentiles • Processing times (scheduled activities), defined by averages or percentile • System throughput – The amount of data passing through a component or system in a given time period or the amount of work performed by a computer within a given time	Mostly automated unless the number of users is minimal. There are various tools available for load testing.

Table 2.6 (Continued)

Test type	Objectives	Test automation
	• Concurrency number of users working with and idle while being logged in the application at the same time • Scalability – The ability to be enlarged to accommodate growing amounts of work or data	
Endurance (soak) testing	To verify that the application can sustain continuous expected loads without performance degradation.	Same as for load testing.
Failover testing	To verify that the system can successfully failover and recover from a variety of hardware, software or network malfunctions without undue loss of user interface, data or data integrity.	This is largely manual testing and coordinated between many other non-functional aspects such as event management, monitoring and alerting.
Resilience testing	To test the ease with which the application handles and recovers from failure scenarios such as power failures, communication failures and software component failure.	This is generally performed manually. However, this can be partially automated. The load testing tools can be used in resilience testing, and AI-generated testing can be used to automate resilience testing.
Accessibility testing	To test that the application or web content is accessible to people with disabilities. It is about building digital content and applications that can be used by a wide range of people, including those with visual, auditory, speech or cognitive disabilities.	Accessibility testing is performed as a combination of manual and automated testing. The following tools are commonly used for accessibility testing: • Dragon Naturally Speaking • Dolphin SuperNova Magnifier and Screen Reader • ZoomText Magnifier/Reader • Job Access With Speech (JAWS) • Read and Write Gold • ClaroView • OpenDyslexic font set

(Continued)

Table 2.6 (Continued)

Test type	Objectives	Test automation
Security tests	To test that the application meets specified security objectives.	Security testing is mostly automated using specific tools and hardware. Product-based companies maintain an in-house security test team to conduct testing. However, customers use external suppliers to perform penetration testing in an SI and multi-supplier environment to ensure that the testing is independent.
Penetration (pen) testing	Pen testing is a type of non-functional testing (white box or black box) that is done to check whether the IT solution is secured or not. Usually, it is an authorised simulated attack on a system that looks for security weaknesses, potentially gaining access to the system's features and data. Pen testing is to verify that the solution meets specified security objectives. This testing may be carried out by a third party.	This is largely automated and performed independently. Some areas of testing can be performed manually, such as user access testing or quality scans. Development teams generally perform manual security testing and vulnerability scanning. A combination of manual testing and automatic security scans reduces potential vulnerabilities, along with penetration testing. Security testing is performed at the end of the development process on the final version of the code that will be implemented into a production environment.
Compatibility testing	To test that the application functions correctly under varying permitted configurations, e.g. different web browsers or user device types.	Compatibility testing is performed as a combination of manual and automated testing using specific tools. The most common areas for compatibility are browsers, operating systems, desktops, mobiles, tablets and mobile networks.
Installation testing	To test that the application can be installed, both as a new application and as a new release, and that it then operates correctly.	This is largely automated testing.

2.4 TRENDS

A trend is a general direction in which something is changing or developing, and in test automation, there are visible trends in increasing demand for tools based on AI, ML and deep learning (DL). In this section, the current trends in test automation tools are covered.

2.4.1 Artificial intelligence, machine learning and deep learning

AI and ML are two upwardly trending fields in recent years, and AI will soon be a tester's best friend, according to the *World Quality Report* 20–21 (Capgemini, 2020). Artificial intelligence enables a machine to simulate human behaviour, and machine learning allows machines to learn from past data without explicit programming. AI and ML are not the same things. At a high level, AI uses ML for knowledge and learning to form intelligent behaviour. AI, ML and DL enable machines to learn, adapt, take decisions and execute through their experience. The intelligence supported by algorithms helps machines to influence the decision-making process and improve over time. Machine learning uses DL for large data management and is about processing large data sets. The DL designs are inspired by human neurons. DL is an AI function that mimics the workings of the human brain in processing data and creating patterns for use in decision making. Many codeless test automation tools use AI to learn from humans, triaging script failures to make tests less flaky.

AI in software development is evolving rapidly, such as in self-driving systems and voice-assisted control. AI-based test automation tools make the testing life cycle easier. There are many testing tools that use an AI-based framework for testing. The AI reduces the direct involvement of the developer or tester in testing. The AI-enabled testing tool reviews the current test status, recent code changes, code coverage and other metrics to decide the test approach, what should be tested and when. This reduces the overall testing effort.

2.4.2 Robotic process automation

RPA is a business automation technology that automates routine, repetitive tasks without human intervention. RPA uses software bots (robots) that interact with the system like humans. RPA is heavily used to automate business processes that are rules-based, structured and repetitive, for example most of the online credit card application processing and personal loan application processing uses RPA to analyse the data, perform a credit check and automatically approve or reject without any human intervention. An RPA 'robot' is a piece of intelligent software to handle repetitive, rule-based digital tasks. A good example of an RPA bot is an internet search engine spider that crawls websites to update search engine indexes. Financial firms were the early adopters of RPA, but there are now companies in many industries, including healthcare, retail and manufacturing, that use RPA technology. RPA systems develop the action list based on end-user interactions on the application's GUI and then perform the automation by repeating those tasks directly in the GUI.

> Robotic process automation is a business automation technology that automates routine, repetitive tasks without human intervention.

There are many tools used for process improvement and these tools fall into two buckets: those that help to investigate and trial possible process improvements, and those that apply the improvement. Process mining tools assist a business process owner to investigate possible process improvements, and RPA/AI tools automate the process. RPA is business process automation and is not business process management. RPA does not provide an end-to-end solution for business management. It uses AI, ML and inputs from multiple sources to make decisions, and it does not process non-electronic and unstructured inputs. RPA provides impressive results for business process automation and offers tremendous value. However, there are many RPA implementation failures. It is common to underestimate the cost and time to implement RPA solutions. RPA provides a tool for testers and business users performing testing, especially in implementing large-scale standard systems like SAP, Oracle and Salesforce.

RPA itself is an automation process. Testing is a component in the RPA life cycle as it needs to be tested and certified that it works as expected. RPA tools are not test automation tools like Selenium or Unified Functional Testing (UFT). For example, Selenium automates the testing of web applications, and RPA automates business processes. Tools such as UiPath, Automation Anywhere and Blue Prism are used for RPA automation.

Many test automation tools are used for RPA automation, for example Selenium scripts can be used for a web-based SUT and can run as RPA in the production environment. However, there are many tools specifically designed for RPA that provide specific features for RPA.

The RPA life cycle involves the following main steps:

- Analysis
- Bot development
- Testing
- Maintenance

RPA requires end-to-end testing to ensure that it works as expected. Generic tools such as UFT and Tosca can be used for RPA automation. 'Record and play' is a suitable and quick solution for RPA test automation. RPA tools are also used in test automation, allowing for codeless automation; they can also be used for regression testing.

The main benefits are:

- Quick regression test pack creation
- Easy scalability
- Business process validation
- Real-time visibility of testing and test progress
- Test on the production environment

To summarise, test automation is a process of automating the testing procedures using a code or software tool such as Selenium or UFT, and RPA is a practice of automating the business process using software robots.

2.5 SUMMARY

In this chapter, we addressed test automation in a variety of IT business models, development processes, and different types and stages of testing. There are a variety of factors to be considered in the decision-making process, as the success of the test automation is dependent on the influencing factors more than the automation tools. The first two chapters will help you to decide if test automation is a viable option to support your overall testing needs. In the next chapter, we will help you choose the right tool for automated testing.

3   TEST AUTOMATION TOOL SELECTION

Introducing test automation to the IT landscape is a management decision, and previous chapters described various factors supporting this decision making. Once the organisation decides to go for automation, the next important activity is automation tool selection. Most widely and commercially used functional testing automation tools are based on object-oriented programming (OOP) or scripting languages. There are many custom-developed tools based on scripting languages such as JavaScript and Shell script. It is important that the tool selection should be in line with the skills available in the team. There are plenty of tools available for test automation, and it is important to select the right tool to automate the solution. The IT market is filled with automation tools, but not every tool suits the requirements and tool vendors tend to offer their tools as the best solution for automation.

This chapter will help you to understand the process of choosing the right automation tool. The first part of the chapter addresses the factors to be considered for tool selection and additional factors to be considered for tool assessment. The second part of this chapter addresses tool selection criteria. A tool evaluation template is included in Appendix E for further reference. The template needs to be customised based on the criteria in this chapter and your specific need. The third part of this chapter covers scripting and coding.

3.1 TEST AUTOMATION TOOL SELECTION

In this section, there is a list of factors to help you in deciding on an automation tool; the tool selection criteria are different for product, project, technology, organisation and project implementation methodologies. All these factors should be considered when selecting a tool for automation needs.

The following generic factors should also be considered:

- Usability of the tool or ease of use
- Functionality applicable
- SUT, infrastructure and tool compatibility
- Compatibility with other project tools, for example in a DevOps environment, the testing tool's ability to be integrated with other development tools
- Maintainability and support of the tool

- Initial and ongoing cost
- Flexibility to adapt to changing requirements

A large retail company in North America reviewed a set of automated testing tools for its web application. The test team decided to use both manual and automated testing as part of the solution to validate the ecommerce application. The tool selection team considered different options fit for both desktop and mobile test automation. Following a detailed analysis of requirements and user distribution, the team selected a mobile automation tool since more than 80 per cent of traffic originated from mobile and tablet devices.

The following criteria should be considered when choosing an automation tool.

3.1.1 Requirements

Requirements are key factors influencing tool selection, and therefore a detailed review of the functional and non-functional requirements is essential before choosing the right tool. Obtaining a deep understanding of the automation requirements, such as solution type (e.g. web, desktop, mobile, on-premises, cloud, hybrid, DevOps, Waterfall, Agile), scope, how testing is to be conducted, post-live test support, percentage of automation, release frequency and the existing team's strength in programming languages, is a key requirement for test automation and tool selection.

A large ERP implementation programme manager decided to procure a tool for its quarterly release regression testing. In the process of tool selection, one of the key requirements was to select an automation test tool to support quick regression testing of the quarterly patch release.

The team reviewed three tools:

1. An open-source tool
2. The tool currently used in the organisation
3. A third-party automation tool recommended by the ERP vendor

The team selected the third-party automation tool recommended by the ERP vendor, even though this tool had the highest annual licence fee, as this was the most suitable tool for their needs. Although the next option was a low-cost, effective tool, this tool required a minimum of 30 per cent changes to the automated test suite prior to the execution of every release. This was not a viable option for the programme.

The outcome of the tool selection process is expected to provide multiple options, including one with 'No Automation' and a ranking of the tool recommendations.

3.1.2 Organisation structure

The organisation structure is an important factor for tool selection, along with the cost and ROI model.

Product development companies have a clear understanding of their requirements for automation, with costs estimated up front and budgets allocated. The tool selection process considers the current products and application in scope and future product lines prior to selection. The technology, programming languages and requirements such as smoke test, regression tests, release cycle and trialling all influence the tool selection.

Organisations that are heavily dependent on external suppliers and have outsourced IT operations have a different approach, and cost management is the main parameter for tool selection. The tool is usually funded by the customer organisation, with third-party suppliers implementing the test automation on behalf of the customer. However, it is likely that the supplier may opt for a low-cost solution if the tool cost and implementation are included in the overall supplier cost. Organisation structure and IT operating models such as fixed price, time and materials, multi-supplier, multi-vendor, product-focused, IT consulting and system integrators influence the tool selection process. The customer organisation needs to set the expectation up front with the suppliers, to provide them with a list of target test automation tools and tool selection criteria. This will deter suppliers from selecting the cheapest test automation tool to reduce their cost over the most suitable tool for the customer organisation.

3.1.3 Development approach

Test automation goals and objectives must be aligned with the IT solution development approach to maximise the benefits. For example, in a DevOps environment, the test automation approach and tool selection need to be aligned with the Agile way of development and support DevOps delivery. The selected tool should be integrated into the solution and the development methodologies used for the implementation. The development schedule plays a major role in test automation tool selection.

3.1.4 Testing objectives

Testing objectives, such as quick regression testing, smoke testing, DevOps, functional automation to support Agile development, UAT coordination and interim patch release, are key influencing factors for appropriate automation tool selection. For example, if the testing objective is to provide a day one TAS to support the DevOps, most likely the team will choose a tool that can be procured easily and require minimum infrastructure to set up and to commence automated testing.

A large retail company in the UK decided to introduce test automation to support its multi-supplier environment. One key objective of the in-house test team was to manage the UAT with a geographically distributed team. After a detailed analysis, the test team selected a cloud-based test management tool to coordinate UAT test case execution and defect management for its globally distributed team. This tool provided a platform for test case execution, defect reporting, tracking, triaging and the management of all the acceptance testing in one place.

3.1.5 System environment and technology

Solution implementation environment and technology are critical factors for tool selection. In recent years many organisations have moved from on-premises to cloud and hybrid environments for solution development, infrastructure and test automation. Testing tool selection should be aligned to the environment and technologies to achieve the expected benefits. A tool that only supports cloud-based solutions may not be suitable for on-premises and hybrid solutions. On-premises requires tools to be downloaded and manually installed by physically or remotely accessing the environment. The infrastructure needs to be provisioned to support the operational aspects of the testing tool, such as access to the environment or periodic patch update. However, a cloud environment may allow an SaaS-based testing tool and periodic automatic patch releases. The testers may be onsite for test automation in a secure environment, and this can be an obstacle for a cloud-based tool supported by an offshore testing team. Testing tools require support from the vendor, known as professional services, and an on-premises client or server environment can incur additional costs over a web-based cloud solution.

3.1.6 Return on investment

ROI is a key factor for tool selection. The ROI and benefits of automation are covered in Chapter 1.

3.1.7 Automation costs

The cost of automation is another factor for tool selection. Some tools have low licence fees but may require high maintenance. A few indicators to identify costs are:

- Implementation effort.
- Licence fee and renewal fee.
- Annual maintenance fee.
- Hardware and hardware support cost.
- Infrastructure cost such as cloud service support fee.
- Hosting platform: cost involved to maintain the platform or the licence cost for the software on the platform.
- Professional services charged by the tool providers.

- Cross-functional support by other teams.
- Script maintenance and enhancement such as ongoing maintenance or enhancement of the script.
- Tool training cost.
- Architecture: the complexity of the system architecture is an indicator of how much effort and cost is likely required to be invested.

Figure 3.1 is a representation of the relative cost of the tools widely available in the market.

Figure 3.1 Automation tools versus cost matrix

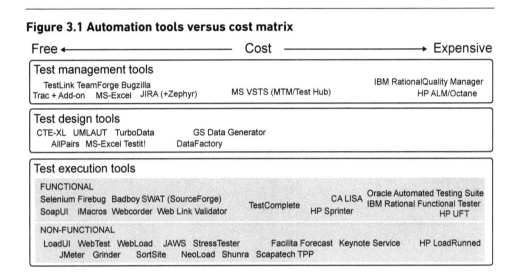

3.1.8 Proof of technology and proof of concept

PoT and PoC are two activities involved in tool selection. PoT is a small project that tests whether an idea about technology is viable, and PoC is a procedure that tests whether a tool fits by analysing it using various parameters. The objectives of PoT and PoC are to verify that the tool can work on different applications and is fit for the organisation's automation needs. It covers the following activities:

- Tool comparison
- Reviewing sample test scripts covering various scenarios
- Automating a few sample scenarios
- Verifying the tool's ability to manage data and support frameworks
- Ability to operate in the IT estate or infrastructure

The PoCs are generally supported by the tool vendors with an evaluation version of the tool.

PoT is a small project that tests whether an idea about technology is viable, and PoC is a procedure that tests a tool's suitability by analysing it on various parameters.

3.2 TOOL SELECTION ASSESSMENT

After collating the list of requirements to be met by your TAS, the next step is to create a comparison matrix to do the tool selection assessment. There are various templates available to support the tool selection process. Tool selection criteria vary based on the organisations' needs and requirements. See Figure 3.2 for the workflow for tool selection.

Figure 3.2 Tool selection process

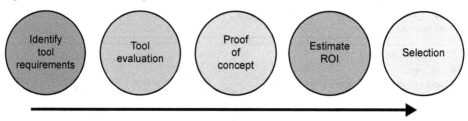

Refer to Appendix E for sample templates on the tool evaluation template. The templates should be customised based on your organisation's needs.

The following factors should be considered when customising the template for your organisation's tool selection assessment:

- Tool usability. Some test automation tools are difficult to learn and hard to use.
- Functions available such as automated reporting, recovery and backups. These features help in saving test automation effort.
- Compatibility with other project tools such as defect management tools and automated build validation. This is essential for DevOps.
- Maintainability such as scripts and frameworks.
- Support from either the vendor or the community.
- Initial and ongoing cost such as tool procurement, renewal, path updates or the cost for the add-on to enhance the features.

- Flexibility to adapt to changing project or product requirements.
- Training, documentation and tutorials available to support the tool and the test automation team.
- Level of programming skills required. More programming will increase the cost and is difficult to maintain the code.
- Level of skills and experience required for automation or maintenance.
- Security and data privacy. For example, security will have to perform a risk assessment of the tool for a secure environment.
- Time available to implement the TAS.
- Supplier's financial solvency, to avoid investing in a tool supplier that may go bankrupt or struggle to keep on developing features.

3.3 CODING OR SCRIPTING

Coding or scripting is one of the essential factors in choosing the right tool for automation. Automated testing and test frameworks are software products, and they inevitably require coding or scripting. Coding enhances the effectiveness of the test automation tool and automated testing.

The following factors need to be considered before choosing the right test automation tool for your needs:

- Programming language used to develop the test automation tool
- Test automation tool compliance with the programming language used to develop the SUT

Coding or scripting can be applied in several different ways, such as in linear, structured, data-driven or OOP forms. This section covers different coding methods to be considered in the tool selection process. Chapter 11 describes the coding method and programming languages in detail for further reference.

3.3.1 Linear scripting

Linear test scripts are created while the user interacts with the application. The test actions mimic the actual user's actions. Scripts are created in a 'record and play' format by capturing the user actions and system responses, recording them in an appropriate scripted format, and re-running them as required. They are linear because those steps through the application are performed sequentially.

3.3.2 Structured scripting

Structured scripting uses two tiers of the script. A high-level script outlines the basic steps and detailed procedural steps that perform the required subroutines. The test tool is programmed with a list of high-level tests to be run. During execution, these will automatically call the required low-level scripts.

3.3.3 Data-driven scripting

Data-driven scripting is where large volumes of data are used from separate data files for testing. The high-level script is created to read data items from the different data files. This reduces the script maintenance required per test.

3.3.4 Object-oriented programming

Object-oriented programming is widely known as OOP and is based on the concept of 'objects', which can contain data and code – data in the form of fields and code in the form of procedures. OOP implements entities like inheritance and polymorphism in programming and binds together data and functions. Python, Java, C++ and Ruby are a few examples of OOP-based languages.

Refer to the 'Useful Websites' at the beginning of this book for external references to find out more about OOP.

3.4 SCRIPTLESS AUTOMATION

Scriptless or codeless test automation is performing test automation without writing any script or using limited scripts. Most of the widely used testing tools and frameworks provide this as a core feature, and this helps business users and manual testers to perform test automation without any or limited knowledge of programming languages. For example, scriptless automation can be used for validating the login functionality with a limited number of user accounts. However, human interaction or programming may be required, in this case, if the requirement is to validate the login for a large number of accounts. The scriptless feature enables testers and business users to automate test cases without coding skill. Codeless automation enables faster results and reduces the automation effort, and is suitable for many requirements. This is a key factor for test automation tool selection as in some cases, a scriptless tool is enough to meet the automation need.

Advantages of scriptless automation are:

- No dependency on programming languages and scripting as there is no coding involved
- More user engagement from non-technical members or business users of the organisation as it requires no or minimum technical skills
- Easy to automate, quick turnaround and payback
- Reduced cost and effort

Disadvantages of scriptless automation are:

- Testing lacks depth and is not thorough as programming enhances the effectiveness of testing.

- No or less scope for customisation as the tools provide limited options to customise without coding.

- One hundred per cent codeless is difficult to achieve.

This book does not cover the details of how to programme or code as there are plenty of books and online resources available for learning coding/programming step-by-step.

3.5 SUMMARY

In this chapter, we addressed the significance of the role of testing tool selection, selection criteria and coding methods to influence the tool selection. Tool selection is an important factor for the success of automation. Knowledge of coding methods is also a key factor for the successful definition of the ideal test automation framework. The coding method will be detailed in Chapter 11. In the next chapter, we will explore another important area: how to build a test automation team.

4   PEOPLE AND TEAM

Test automation requires a highly skilled team, and understanding how to build a successful automation team is crucial. This chapter is about building and leading a highly successful test automation team.

> Building and leading a successful test automation team requires a careful definition of team structure, role descriptions, thorough selection of team members, extensive training, a well-defined career path and development plans.

4.1 TEST AUTOMATION TEAM

The test automation team is expected to have the right mix of programming skills, soft skills, attitude and aptitude. The key factors you should consider in building a test automation team are explained in this section.

4.1.1 Importance of having the right skills

Test automation is often considered as a solution to replace manual testing, manual testers and perform everything with a mouse click. The one reason for this is the sales pitch from automation tool vendors that claim test automation is akin to 'record and playback' or 'scriptless' automation. The truth is that any sustainable and long-lasting test automation requires a realistic fact-check of the constraints involved in building an optimal test automation team. This follows many of the same disciplines as developing any successful team. However, building an automation team is more difficult than building a software engineering team, as the test automation engineers should have multiple skills, such as coding knowledge, testing skills and good business sense. The automation team consists of experts who know how to develop an automation framework, understand software testing and have good technical skills. If seeking expertise from external sources to build the automation team and framework, the leadership should consider bringing on board a professional testing service company to do this job.

The professional testing services partner will develop the test automation team and framework along with consideration of what is needed in the long run. While a professional testing services company increases the set-up cost, their expertise is invaluable in the important questions they raise up front, including sustainability, in-house test maturity and tooling considerations.

4.1.2 Team and roles

A test manager or Scrum master should be able to build and lead a high performing test automation team that provides a return on investment to the organisation.

The success of any team is heavily dependent on the leadership skills of the team lead; in other words, a team will inherit and follow the leader's attributes. The team lead is the most important hire you will make when building your test automation team because the leader sets the tone. The team lead should have a good track record of building multiple diverse test automation teams with sound experience in hiring the right resources.

The team certainly needs people with good experience in developing frameworks and good coding skills. Team members should be interested in learning and supporting the development of the automation framework. The latter part of this chapter covers new member selection and interview criteria, their previous experience and their value addition to the team. The interview focus should rely on their previous work, hands-on experience and automation knowledge.

A large public sector application decided to build a test automation team to increase efficiency in their testing. The testing had grown exponentially over the last few years, and the test team doubled its size within a year to manage the additional work. A large amount of testing time was required for data validation and regression testing. The SUT was developed on Microsoft technologies, and most of the solution was hosted in an on-premises environment.

They hired an automation test lead with excellent experience in building a wide variety of automation frameworks, programming skills in an object-oriented language, good team leading skills, tool evaluation skills and manual testing experience. Within a few months, the automation test lead managed to build a team, create a framework and automate data validation testing. The team managed to meet the ROI within six months.

Test automation is a platform of constant change and improvements, and the team members should be open to this. Willingness to learn new technology and tools is essential. The initial members of the team should be a group of specialists such as an SME on the vendor tool, an expert in the specific programming language, a seasoned test automation framework architect and a test manager. The team structure can be defined based on the need for the roles outlined below:

- **Automation test manager.** This person facilitates and manages team governance. This role is equivalent to a test manager or a Scrum master. The responsibility involves reporting, analysis, test management, tool acquisition, governance and administration, and the skills they bring on board are the ability to have difficult conversations and negotiations, and stakeholder management with a sound automation background.

- **Automation test team lead.** This person builds the automation framework and should be excellent in mentoring others on how to use and extend the framework. This person must be skilled in test automation, documentation and programming languages.

- **Automation architect.** This role can be part of the automation test team lead or a separate one, subject to the amount of work to be done and structure of the team. This person is responsible for designing the framework and requires all the attributes of a successful solution architect.

- **Test automation engineer.** This person supports the lead developer in building the framework. This person should be good at maintaining the framework and automating the SUT using this framework.

4.2 SKILLS

This section explains the important skills required for test automation engineers. There are three ways by which professionals move to test automation:

- Manual testers shift to automation.
- Engineers specialise in test automation.
- Developers shift to automation.

The key prerequisites of a good test automation engineer are covered below.

4.2.1 Generic skills

Generic testing skills (also known as key skills or essential skills) are expected from all the members of the testing team. These skills involve interpersonal as well as testing skills.

Interpersonal skills are the behaviours and tactics a person uses to interact with others, or the ability of a person to work well within the team and with other teams. Interpersonal skills consist of communication, troubleshooting attitude, listening and so on. They are essential for more collaborative teams like DevOps or Agile.

Test certifications such as ISTQB Foundation are a good method to ensure a person has generic testing skills. An understanding and sound knowledge of manual testing concepts and experience is essential for a test automation engineer; after all, it is a testing job! Remember not all testing can be automated, and automated testing is not a replacement for manual testing.

One of the key skills for any tester is to build a good level of knowledge of the SUT, and test automation engineers are not excluded from this. Test automation engineers need to understand the software SUT inside and out.

Test automation engineers should have a good understanding of the following generic testing skills, testing areas and concepts:

- Automation frameworks
- Test architecture
- Test strategy and plan
- Test design
- Non-functional testing
- Configuration management
- Test environments
- Release and deployment
- SDLC types and their characteristics
- Application functional knowledge
- Domain knowledge such as banking, HR, travel or transport
- Business-critical features
- Manual testing processes and techniques
- Technology such as cloud, on-premises, legacy, web or client–server
- Programming languages and packages used in the SUT
- Underlying technologies such as ERP or database
- Automation and market trends
- Converting business requirements to testing requirements
- Interfaces and technologies such as web services and application programming interface (API)
- Quality models such as Test Maturity Model integration (TMMi) and International Organization for Standardization (ISO)
- Risk and issue management
- Project priorities and timeframe
- Good communication and presentation skills

4.2.2 Test automation skills

Expertise in test automation, automated testing and programming are essential for a test automation engineer. These enable them to build high-quality test frameworks and scripts for automated testing. The appropriate assertions, that is coding standards, must be built into the scripts to ensure those automated tests are high quality. Below are some core skills required for test automation engineers:

- Knowledge of test automation tools, such as Selenium, UFT and Eggplant.
- Expertise in different automation frameworks (automation frameworks are covered in Chapter 5).
- Analytical, problem-solving and logical skills.
- Technical skills, such as:
  - coding
  - scripting
  - object-oriented programming
  - automation architecture
  - automation concepts
  - test data management
  - security standards
  - exception handling
  - reporting and tracking
  - libraries
  - variables
  - test script management
  - programming or scripting languages, such as Python, Pearl, JavaScript, Structured Query Language (SQL) and REST
  - coding standards

4.2.3 Competency matrix

Table 4.1 displays the main competency requirements for automated testing roles.

Table 4.1 Competency matrix for test automation

Skills required	Test automation engineer	Test lead/ architect	Automation test manager
Analytical and logical skills	High	High	High
Test planning and estimation	Low	High	High
Test management	Low	High	High
Stakeholder management	Medium	High	High
Client or user awareness	Medium	High	High
Hiring and team building	Low	Medium	High

(Continued)

Table 4.1 (Continued)

Skills required	Test automation engineer	Test lead/ architect	Automation test manager
Team leading and management	Low	High	High
Communication and interpersonal skills	High	High	High
Presentation skills	Low	High	High
Domain knowledge	High	Medium	Medium
Defect management and tracking	High	High	High
Experience of using testing tools	High	High	Medium
Database	High	High	Medium
Operating systems and browsers	High	High	Medium
Mobile platforms	High	High	Medium
Manual testing	High	High	High
System testing	High	High	High
Integration testing	High	High	High
Non-functional testing	Medium	High	Medium
User acceptance testing	Low	Medium	High
Test design and development	High	Medium	High
Test execution experience	High	Medium	High
Test scripting	High	Medium	High
Testing methodologies	Medium	High	High
System development and tools	High	High	High
SDLC and software test life cycle (STLC)	High	High	High
System architecture	Medium	High	Medium
Development tools	High	Medium	High
Configuration management tools	Medium	High	Medium
Technical knowledge	High	High	Medium

4.3 HIRING PROCESS

Testing, as a profession, implies quality and attention to detail. Attracting the right talent and professionals with a testing mindset is half the war won. As a test automation leader, you must ensure the corners are not cut when it comes to selecting the right team. The first step to hiring the right automation resource is to define a job description for the role.

4.3.1 Job description

This section covers what the job description, widely known as JD, is and defines the skills that a test automation engineer should have. A JD describes the roles, skills and responsibilities expected for the position. Automated testing is a highly technically focused role, and central to the role is the hands-on development and execution of automated testing throughout the complete life cycle, from project initiation through to delivery. A JD includes:

- Description of the organisation
- Role description: responsibilities, essential and desirable skills

This section gives a guideline to the skills that are required for a generic automation role. The role is based upon functions performed in the team, and skill is acquired through further training and experience. Roles in the team are relatively constant, and skills develop and change rapidly based on the nature of the automation, application and customers.

A sample set of generic, essential and desirable skills for a test automation engineer are listed below.

Generic skills and responsibilities:

- An enthusiastic, highly motivated and methodical individual
- Sound knowledge of software solutions
- Excellent team and stakeholder communication
- Ability to work on an individual basis and as part of a team
- Good test and time management
- Reporting progress to the test manager, Scrum master or project manager

Essential skills:

- Experience in automation frameworks, automation scripts and execution of scripts
- Proven experience of test automation development using tools such as the following:
  - Automation tools, for example Selenium, UFT and Eggplant
  - Static analysis tools, for example Checkstyle, FindBugs and SonarQube
  - Unit test, for example JUnit and JavaScript
  - Performance or load test, for example LoadRunner and JMeter
- Good working knowledge of SDLC tools such as:
  - Version control, for example Git and Stash
  - Collaboration, for example Confluence and Wiki
  - Defect management, for example Jira, Azure DevOps and application life cycle management (ALM)

- Good working knowledge of SDLC tools such as:
  - Azure DevOps, Micro Focus, ALM and TeamForge
  - Version control: Subversion, ClearCase and Azure DevOps
  - Collaboration: Confluence, Wiki and SharePoint
- Sound knowledge of product and solution testing and quality assurance
- Good knowledge of software development and software testing life cycles
- Proven experience in using SQL Server and writing advanced SQL queries
- Experience in installing and configuring virtual environments
- Hold or be actively working towards ISTQB Advanced Level Technical Test Analyst

Desirable skills:

- Experience with API testing tools
- Experience in browser, mobile and desktop-based applications
- Experience in distributed software architectures and cloud-based solutions
- Able to create and interpret architecture diagrams
- Experience in environment, release and deployment management

4.3.2 Curriculum vitae review

The next important step for hiring a test automation engineer is the CV review. The following section covers a few questions to be considered in a CV review:

- What is the candidate's experience in automation, automation framework and programming languages?
- Have they included a broad range of relevant experience?
- Have they used examples to demonstrate these skills?
- Do they follow trends in test automation? Do their projects follow the tool journey, such as a different version of the tool? For example WinRunner to QuickTest Professional (QTP) to UFT, or the 30-year tool journey from Microsoft Windows 1 (year 1985) to Windows 10 (year 2015).
- Do they have a LinkedIn account? Check the information on the account and whether it is in line with their work and networks.

Refer to Appendix D for sample skill set of test automation engineers.

4.3.3 The interview

The interview structure is heavily dependent on the company and its culture. Most companies carry out an initial screening by the human resource team to check for company and role fit, followed by a short telephone technical interview. Many companies have moved the interview process to virtual platforms. A video interview is now considered as close to an in-person meeting in understanding the candidate and their responses.

This section covers a standard interview structure. The first stage is an initial technical screening.

Interview 1: initial screening (approximately 30 minutes):

1. Introduction
2. Testing and automation capability interview based on CV
3. Summary of the interview and the next steps

The second stage is a detailed technical screening based on a test automation-related case study.

Interview 2: technical assessment (approximately 120 minutes):

1. Introduction
2. Case study based on the approach to automate or create an automation framework, for example an application such as a credit card form or social media account
3. Candidate presentation of the case study
4. Interview based on the case study
5. Technical interview based on CV
6. Question and answer (Q&A)
7. Closing and next steps

The presentation relates to the candidate's understanding of automation, framework development and test life cycles. The presentation will be based on the case study, and will greatly depend on the candidate's relevant business, tools, automation, industry, companies, products, location and technology experience. The other aspects that may be verified in the interview are shown in Table 4.2.

In most companies, this interview is performed by a senior person, such as a group head or a head of a product line, along with the human resources team to check for the match of the role and the company.

Interview and final assessment (approximately 30 minutes):

1. Introduction
2. Generic interview

Table 4.2 Skill assessment matrix

Testing types	Test management
• Functional	• Exit and entry criteria
• Non-functional and operational	• Reporting
• Interfaces and integration	• Defect management
• APIs and messaging	• Environments
• Failure and recovery	• Release
• Regression	• Testing life cycle
• Business processes	

Test objectives	Test automation
• Testing timescales	• Tool selection
• Test coverage and traceabiity	• Automation framework
• Test responsibilities	• Automation test strategy
• Deliverables	• Scripting language
• Assumptions and constraints	• ROI
• Risk and issues	

Refer to Appendix C for sample interview questions.

4.4 SUMMARY

In this chapter, we explored the 'people' aspects of test automation and the skills required by the team. We also addressed the competency matrix and hiring process. Next, we will explore another important area: the test automation framework.

5   AUTOMATION FRAMEWORKS

This chapter is essential for all senior test automation professionals – such as managers, leaders, project managers and other decision makers of the automation framework. It details the test automation framework, widely used frameworks, and the advantages and disadvantages of each. This chapter also addresses how the frameworks manage data and the philosophy of using function libraries of automated test frameworks.

A test automation framework is a set of rules followed in a systematic way to deliver targeted results. It is a platform developed by integrating various hardware and software components, along with using various tools for test automation. It is an integrated solution of function libraries, test data, object details and various reusable modules.

A test automation framework is a programming framework that consists of a comprehensive set of guidelines to produce beneficial results from the automated testing activity.

In general, a test automation framework is created through the use of programming languages such as Python, Pearl, Java, .NET and Shell script or testing tools such as Selenium, UFT, Tosca and Eggplant to fulfil a specific need.

The framework should be fit for purpose, support the current and future products and applications in the infrastructure estate, and independent of programming languages, if possible, to support the team in building knowledge and expertise quickly.

The benefits of the test automation framework are:

- Efficient automated test script development
- Provides a structured development methodology to ensure the uniformity of design across multiple test scripts to reduce dependency on individual test cases
- Reuse of components and code

- Reduces dependence on teams by automatically selecting the test to execute according to test scenarios
- Improves the utilisation of various resources
- Ensures an uninterrupted automated testing

The key features of any framework are:

- Runs more tests per cycle.
- Minimal manual intervention.
- Returns quicker results.
- Reusability of codes and components.
- Enhances efficiency of script development.
- Provides a structured development methodology.
- Reduces dependency on individual tests.
- Ensures uninterrupted automated testing.
- Low cost maintainability.
- Improved failure recovery and exception handling.
- Automated test execution dashboard and reports.
- Optimal error handling process.
- Test data remain independent of the code.

The most widely known test automation frameworks are:

- Linear testing framework
- Modular testing framework
- Library architecture testing framework
- Data-driven testing framework
- Keyword-driven testing framework
- Hybrid testing framework
- TDD testing framework
- Behaviour-driven development (BDD) testing framework

This chapter describes the various test automation frameworks in detail.

5.1 LINEAR TEST AUTOMATION FRAMEWORK

The linear test automation framework is generally known as the record-and-playback framework, and it works in linear or sequential mode. Tests and test scripts are created individually and executed one after the other.

Figure 5.1 Test automation frameworks

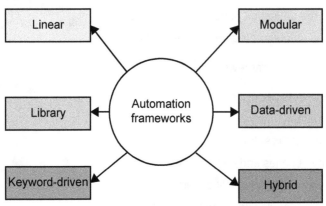

In linear scripting, the test actions mimic the actual user actions that would be performed while using the application. The scripts can be written from scratch and they are usually implemented by a tool that provides a 'capture and replay' facility. This operates by capturing the user actions and system responses and recording them in an appropriate scripted format (see Figure 5.2). The scripts can then be re-run as required.

A leading global active asset manager and investment firm revamped a core asset-management discipline. Testing was a key activity in the SDLC, and test automation was a key solution for continuous releases. The test team found it difficult to automate the functionality due to a lack of domain knowledge. The business analysts and product owners used record and playback to create the initial automation suite. This was a starting point for the test automation team, and the team enhanced the scripts and converted them to a competent linear automation framework for the SUT.

Advantages:

- Simple and does not require advanced technical knowledge.
- Does not require writing code.
- Generates test scripts quickly.
- Well-suited to create a proof of concept.
- Automation can be introduced very quickly.
- No programming skills are required since it is click and record.
- Linear framework requires no technical or programming knowledge.
- Quick to create a linear framework.

Figure 5.2 Linear test automation framework

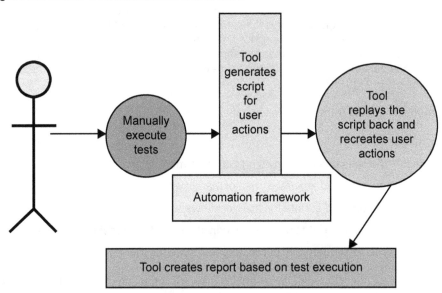

- Low learning curve.
- No initial development or learning effort.

Disadvantages:

- Not reusable and requires high maintenance.
- Test data are hard-coded and difficult to maintain.
- Limited coverage on test data.
- Lacks integration between test cases.
- Difficult to run a large set of test cases.
- No effective error and exception handling.
- High manual intervention.
- The script is only applicable to the recorded set of actions.
- Changes to the SUT results in the updating of numerous recorded scripts.
- Change in application workflow requires high rework.
- Lack of scalability.
- The scripts capture all the user actions, including any incorrect ones.
- The script may become inefficient due to repetitive user actions (e.g. repeated browser back button usage).

5.2 MODULAR TESTING FRAMEWORK

The key feature of the modular testing framework is to divide the system under test into stand-alone and logical 'modules' based on the OOP concept known as 'abstraction'. Independent test cases are developed for these modules and together build a larger test suite for the solution. These modules are separated by an abstraction layer, and changes made to one part of the solution or the framework has a limited impact on the overall framework (see Figure 5.3).

The modular testing framework follows these steps:

1. Analyse the SUT

2. Identify the reusable workflows and develop manual test scripts

3. Create common and separate functions and units

4. Develop automated scripts for the workflows

5. Create a master script to invoke the individual and functional test scripts

6. Combine tests sequentially to create a test suite for the modular framework

The key approach for the modular testing framework is the use of the abstraction layer, and this helps to avoid changing the overall framework whenever there are changes to individual sections. The corresponding module and the associated individual test script need to be changed if there are changes made to the application.

Figure 5.3 Modular test automation framework

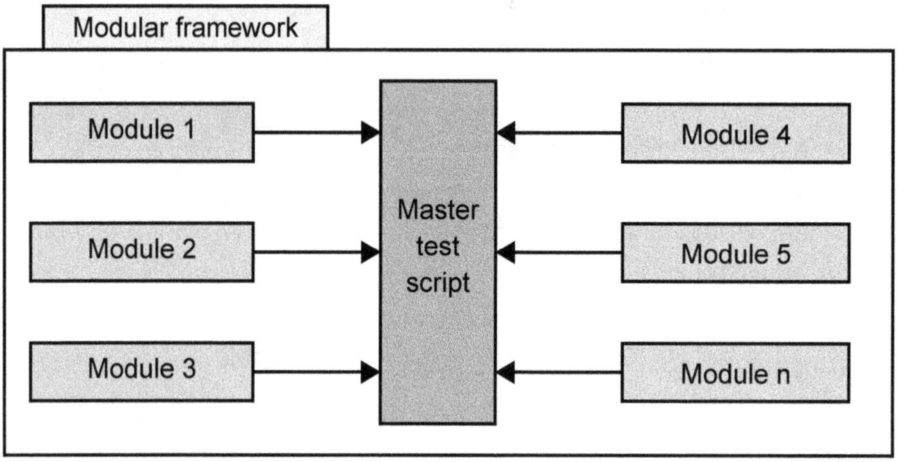

Advantages:

- Low maintenance
- Well-structured modules
- Ease of use
- Less effort to create test scripts
- Reusability
- Easy reporting and debugging
- Scalability
- High team productivity, efficiency and accuracy
- Quick ROI

Disadvantages:

- Data are hard-coded and embedded.
- Difficult to use different data sets.
- Basic framework and is time-consuming to maintain the framework.
- High maintenance required.
- Programming skills are required to develop the framework.

5.3 LIBRARY ARCHITECTURE TESTING FRAMEWORK

The library architecture testing framework is an extended and enhanced version of the modular testing framework. The framework is built based on the library approach, which involves setting up common functions under a library and accessing those functions in the test scripts. This is a more structured and maintainable framework compared to the modular and linear frameworks. Library architecture is more intelligent and organised compared to a generic modular framework (see Figure 5.4).

Advantages:

- Developed on modular framework architecture
- Overall low maintenance
- Scalability
- Cost-effectiveness
- Reusability

Disadvantages:

- Abstraction maintenance is difficult.
- Good technical knowledge required.

Figure 5.4 Library architecture test automation framework

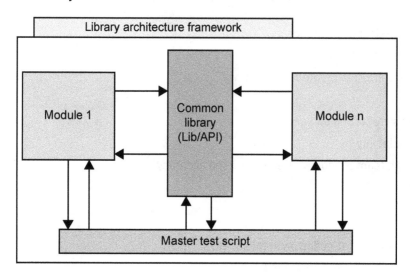

- Test data maintenance is complex.
- Highly technical.
- Test script development is time-consuming.

5.4 DATA-DRIVEN TESTING FRAMEWORK

The data-driven framework is a testing framework in which input values are read from data files and stored into variables in test scripts. The data-driven framework and methodology are used to automate and validate applications that require large volumes of test data and swift testing. This framework helps to create test data, calculations, test iterations and permutations with minimal effort. This helps testers validate the solution with different data sets, including positive and negative tests, into a single file and test script.

The driver scripts contain navigation through the program, reading the files and logging the test status information (see Figure 5.5).

The most notable feature of the data-driven framework is its ability to separate test data from code. This allows data to be stored on an external data source to enhance data integrity and security. This helps to use different data in the different testing cycles and supports millions of different data combinations. The main difference and advantage of the data-driven framework over the linear or modular testing frameworks is that the test data are not hard-coded. The data-driven test framework can store and pass the input data from an external data source, such as spreadsheets, text files, comma separated value (CSV) files and databases. Test scenarios can be extended to numerous

Figure 5.5 Data-driven test automation framework

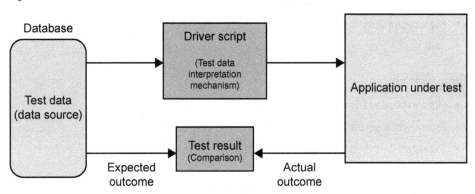

test cases with minimal effort. Test data can be sourced from a predefined spreadsheet, and a business analyst or SME with no knowledge of the framework can prepare the test data for the automated testing.

The framework has the capability to access real-time data from production or live source and validate the solution with the most up-to-date and transactional data without losing data integrity and security. This is ideal for industries such as retail, banking and healthcare where the application requires large data sets for validation.

A large global financial application used the data-driven framework to providing 100 per cent test coverage to validate millions of user accounts. Data validation testing was a key priority for the testing, and data involved General Data Protection Regulation (GDPR) and personal identifiable information (PII). Manual testing was not feasible as the data set consisted of millions of interrelated data. There were other challenges to the testing as the test environment was not accredited to store or process live-like PII data.

A data-driven framework was effectively used in this financial application to validate the testing of millions of customers' data against the production database as part of legal compliance. The data-driven framework was designed and installed on the live environment in such a way that it could access the production database securely and validate data with 'read only' permission. This helped the application test all the data against the latest database without any security breaches.

Advantages:

- Test multiple scenarios with fewer lines of codes
- Quick testing of multiple data sets
- Secure test data management

- No test data hard-coding
- Easy test scripts maintenance
- High reusability
- Test script independent of test data

Disadvantages:

- High expertise in programming skills required.
- Framework setup requires significant time and effort.
- Requires continuous data maintenance.

5.5 KEYWORD-DRIVEN TESTING FRAMEWORK

The keyword-driven framework uses a technique to separate keywords or actions and data from the test scripts. The keywords are stored in an 'object repository' to allow for better code reusability and less script maintenance. The function of the SUT is stored in a file, usually in a table format with instructions. A keyword-driven framework is a table-driven or action-based testing method. The keywords represent various actions being performed to the application.

The fundamental approach of the keyword-driven framework is to separate the tests into smaller components such as test steps, object and actions. Data are managed separately, and expert knowledge is required to develop the framework. This framework and keywords are also used to run the tests manually (see Figure 5.6).

Figure 5.6 Keyword-driven test automation framework

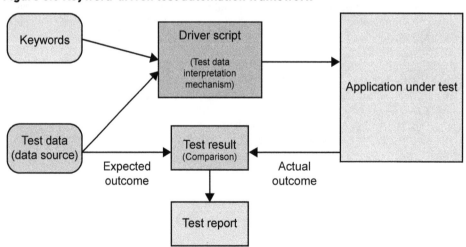

5.5.1 Keyword table

The keyword-driven framework is used to speed up automated testing by utilising the keywords for a common set of actions. The framework separates the coding from the test case, test step and data in a table, and this helps a non-technical person to update the keyword, create the data and run the existing framework.

The keywords are stored systematically with an associated object and actions in the object repository.

- Test step description – The action to be performed
- Test object – The name of the web page Document Object Model (DOM) or element, such as Username and Password
- Test data – Such as username and password
- Action – The name of the action such as 'click' or 'select'

The test automation engineers refer to Table 5.1 and create test scripts. Test scripts use the keyword from the table for the corresponding actions. During the test execution, the test data are used from the data source (Table 5.1) for the respective username and password field, executing the relevant script. This avoids using hard-coded test data, specifically sensitive passwords in the script.

Table 5.1 Example keyword table

Steps	Test step description	Test objects	Test data	Actions
Step 1	Log on to the social media site	Gologin	URL	Click Button
Step 2	Enter user ID/userid	UserID	Input data	Enter
Step 3	Enter password	Password	Input data	Enter
Step 4	Click on Log In button	Gologin	NA	Click Button

Advantages:

- Minimal scripting knowledge is required to automate new features.
- Keywords are reusable for new tests.
- Low maintenance in the long run.
- Concise test cases.
- Early creation of the framework without SUT.
- Independent of test scripts, keywords and test data.
- Easy comprehension by a non-technical audience.

Disadvantages:

- High technical and programming skills required.
- High development cost.
- Complex and requires detailed design and approach.
- Keywords, object repositories or libraries have to be created up front.
- Continuous maintenance.

5.6 HYBRID TESTING FRAMEWORK

The hybrid testing framework is a combination of one or more frameworks. Previously, the hybrid test automation framework widely used the concepts of keyword- and data-driven frameworks. However, the modern-day hybrid frameworks use the best concepts from various frameworks to maximise the benefits (see Figure 5.7).

Figure 5.7 Hybrid test automation framework

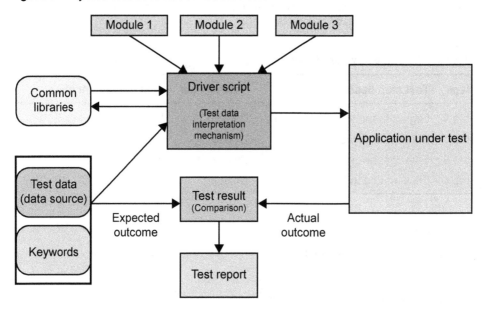

The key to developing a hybrid testing automation framework is choosing the right tool for it. It is important that the framework adapts to the SUT and the other variables involved in its delivery. Most frameworks turn out to be hybrid over a period. Hybrid frameworks that use the concepts of keyword-driven and data-driven approaches tend to be more successful in test automation.

Advantages:

- Fit for various applications.
- High adaptability.
- Granularity of test reporting.
- Extensibility.
- Best concepts from various frameworks.
- Accurate and highly efficient.
- Test data, functions, tests and libraries, for example, are separated.
- Reusability of code.
- Good for testing large data sets and complex applications.
- High ROI over a period.
- Good test coverage.
- Improved failure recovery and exception handling.
- Minimal manual intervention.
- Detailed dashboard.

Disadvantages:

- High and continuous maintenance.
- Considerable technical knowledge and expertise in the scripting language is required.
- Time-consuming.
- Requires detailed test plan and architecture.
- High learning curve.

5.7 TEST-DRIVEN DEVELOPMENT TESTING FRAMEWORK

TDD is generally considered a development technique and unit testing approach. However, TDD is widely used for DevOps. Hence, it is generally used along with the automation frameworks mentioned in the previous sections.

The main concept of TDD is to write unit tests first before you develop the code. The test fails the first time round as there is no corresponding functionality that is developed. TDD develops better software and is primarily driven by predefined tests, thereby increasing the speed of tests and accuracy (see Figure 5.8).

Advantages:

- Defines acceptance criteria up front.
- Reduces the risk of interpreting the expected behaviour.

Figure 5.8 Test-driven development

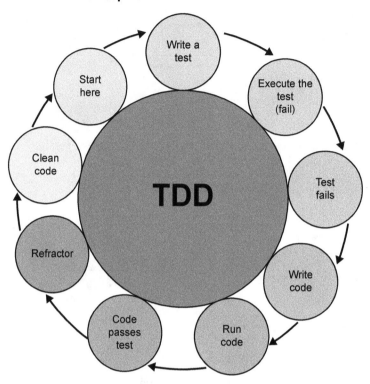

- Focused and productive development team.
- Higher code quality.
- Follows good coding practices and guidelines.
- Creates an automated regression test suite.
- Speeds up the development productivity in the long term.
- Clarifies dependencies much earlier in the development cycle.
- Reduces code and requirements-related defects.
- TDD levels up the quality.
- Tests are usually written for different scenarios.
- High ROI.
- Detailed and documented unit testing.
- Follows the approach 'test first and develop the SUT later'.
- TDD results in more tests.
- Better code coverage.

- Reduces the percentage of bugs.
- The quality of the product developed with TDD is significantly higher.
- Documentation can be used further on for any purposes.

Disadvantages:

- Errors in TDD tests cases will increase defects.
- High up-front time and effort.
- Initially slows down the development.
- High effort to maintain the test suite.
- Continuous housekeeping.
- Unit tests are hard to write up front.
- Hard to develop across all environments, such as legacy systems.
- Highly misunderstood and referred to as 'functional testing'.
- Takes time to make these tests.
- Developers can focus on the test more than the actual functionality and code itself.
- Lacks code optimisation.
- Increases the development effort.

5.8 BEHAVIOUR-DRIVEN DEVELOPMENT TESTING FRAMEWORK

BDD (see Figure 5.9) is considered an extension of TDD as the process is designed to enhance the delivery of software development projects by improving the communication between engineers and business professionals. BDD ensures all projects remain focused on delivering what the business needs while meeting all the requirements of the end-user.

BDD is suitable where the business owner is familiar with the unit test framework. In BDD, test cases are written in a common language used by the business. BDD enhances collaboration between the technical and business teams.

BDD scenarios focus on the expected behaviours of the target application. This helps the project team to work collaboratively and ensures they are on the same page. The BDD scenarios are used as requirements, acceptance criteria, tests and candidates for test automation. BDD frameworks enhance test automation as it directs the test automation engineer to develop a method or function for the corresponding actions.

Cucumber is a widely known test automation framework for BDD, and in Cucumber the BDD specifications are written in plain English using Gherkin language.

Figure 5.9 Behaviour-driven development

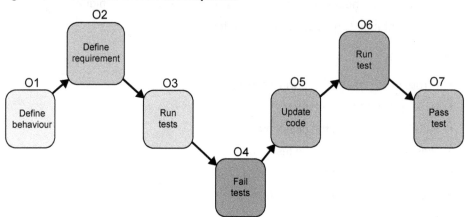

Advantages:

- Strong interactions between the project team and the business.
- BDD increases and improves collaboration.
- Software design follows the business value.
- Solution development meets the user needs by focusing on the business needs.
- Early discovery of the unknowns.

Disadvantages:

- Time-consuming
- More effort in writing automation code for keeping that code in sync with the steps defined in the BDD scenarios
- Difficult to write automated tests up front

Behaviour-driven development is generally used alongside the acceptance test-driven development. ATDD is developed based on the end-users' views. BDD focuses on the application behaviour, and ATDD emphasises the business requirements. In TDD, BDD and ATDD tests are written before the implementation takes place. TDD is a development practice, while BDD is a team methodology and ATDD is a user acceptance methodology.

5.9 SUMMARY

In this important chapter, we addressed what a test automation framework is and walked through the widely and commonly used types of test automation frameworks, including the advantages and disadvantages of each. In the next chapter, we will cover another important aspect of test automation, the test environment.

6   ENVIRONMENTS

Environments play an essential role in the success of test automation. This chapter covers various IT infrastructure and environments and their role in automated testing.

The test automation framework and automated test suites are designed to work independently of environments. An environment-specific automated test suite will limit the opportunity to reuse the automation suite.

A test environment is a collection of software and hardware in which the testing takes place, and is generally a cut-down version of a production environment. An integrated test environment is a 'federated' environment that is representative of the end-to-end solution to prove that the proposed solutions work together.

Test automation needs to verify the test environment is working as expected. An ideal automation test suite will be independent of environments, platforms, browsers and mobile devices, that is, cross-platform compliance. Maintaining a separate automation suite for different environments is not feasible and is effort-intensive. Most test automation tools offer cross-platform compliance.

A large US-based product development company used automated testing for release and regression testing. The company has the same product lines in different platforms such as Windows, macOS, Linux, IRIX and Unix, along with multiple browsers and mobile versions.

The products often go through changes, patch updates and hotfixes. Regression and release testing were a challenge to perform on all the platforms and devices.

After a detailed study, the test automation team developed a custom framework to support cross-platform testing. The automated testing followed a data-driven automation suite for release and regression testing. A spreadsheet was used to manage cross-platform testing, a simple data sheet that listed all the different platforms and versions with scope and out-of-scope sections. Based on the scope value the automation suite included a selected platform for testing.

The automation team anticipated cross-platform requirements and developed libraries to manage to test for all different versions and platforms. The test execution was managed from a simple spreadsheet with 'yes' or 'no' values.

6.1 TEST AUTOMATION ENVIRONMENTS

The technical and security specifications of different environments are a challenge when designing and developing the automation suite. The test automation approach should be designed in line with the DevOps or cutover and release approach of the solution to ensure that the test automation suite and the automation framework function independently in different environments. Test automation is expected to work on various environments and infrastructures with minimum changes.

A test environment is a set of software, devices and infrastructure in which the testing team tests the SUT.

The major environments are:

- Development
- Test
- Production support
- Production

Each of these environments may consist of multiple sub-environments, such as system testing, system integration testing, end-to-end integration testing and performance testing (see Figure 6.1). Building and managing test environments efficiently and effectively in line with the test automation strategy helps to achieve successful test automation outcomes. This can deliver significant benefits and substantial cost savings for the organisation.

Table 6.1 shows different environments and their features.

A key principle in the provision of the automation test environments is that they should closely resemble the architecture, configuration and release process of the target live or production environment, for example if the live environment is hosted on physical servers (e.g. on premises) then testing using environments hosted on virtual servers (e.g. cloud) can introduce additional risk.

6.1.1 Development

The development environment is where development takes place. This can vary from a developer's desktop to a fully configured and version-controlled system. A development environment is a place where most of the unit test automation preparation takes place.

Figure 6.1 Test environment sequential diagram

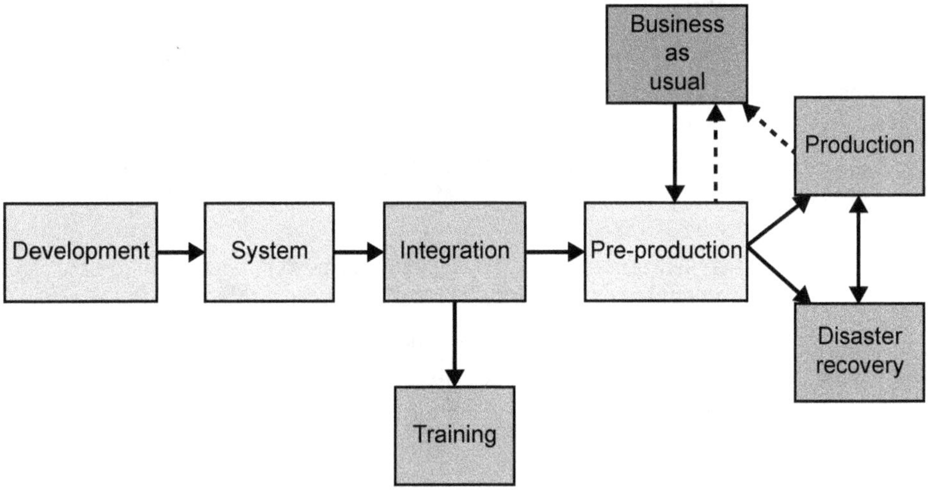

Table 6.1 Environments and their features

Environment	Features
Development	Product development is conducted in the development environment.
	Used for automation test preparation.
	Generally, conducted outside any configuration management processes to allow for a 'DevOps'-like delivery approach.
	Used for compiling build and proving release items.
	Emulation models and test stubs used for inbound and outbound components.
	Development or infrastructure team is responsible for this environment.
Testing or staging	Testing takes place in the test environment.
	Automation test execution takes place in this environment.
	Release and deployment are configuration managed.
	Generally owned by the testing team.
	Automated regression suite is used in this environment to prove the build stability.
	Emulation models and test stubs are used for inbound and outbound components as required.

(Continued)

Table 6.1 (Continued)

Environment	Features
Pre-production or business as usual	The pre-production environment is generally an informal environment used to verify the connectivity with existing infrastructure and applications.
	The pre-production environment is generally a fully integrated environment to prove the end-to-end connectivity between source and target systems.
	This environment is generally used for performance testing and early application security testing.
	The pre-production environment is generally used for UAT and live support.
	Automated test suite is used for regression testing.
Training	The training environment is a 'nice to have' environment to allow UAT and training activities to be performed independent of each other.
	This environment is owned by the business or the product team.
Production/ live	Release and deployments are configuration managed.
	Jointly owned by the IT, business and product teams.

This environment is hosted within the internal network, and not all infrastructure or platform releases may be applied.

The benefits of using the development environment for test automation are:

- Early automation is possible as it is the first environment in line.
- Test automation can start along with the development.
- Test automation can be used to build validation and smoke testing.

The challenges of using the development environment for test automation are:

- Unstable application and constant code and infrastructure changes. A stable SUT is recommended for automated functional and regression test preparation.
- The development environment is not a true representation of the production environment for automation.
- The automated test pack may require changes in other environments prior to execution.
- Stubs and drivers are generally used to develop the SUT and automated test scripts, and this may impact the quality of the automation test suite.
- Regular backup for the automation test scripts may not be possible.

6.1.2 Test

The test environments cater to the test team, allowing them to verify the solution functions according to the specified requirements. The test environment can be stand-alone or integrated. The integration test environment is used to test the inter-process communications to ensure that the system parts function together correctly. The cloud infrastructure provides the opportunity to spin off any environment quickly, in almost the same way as the production environment. Most cloud providers specify clear guidance for code development and release.

The common test environments are:

- System testing
- System integration testing
- End-to-end integration testing
- Non-functional testing

The test environment is the right place to commence test automation, as it is built for testing, configuration management and release control.

The benefits of using the test environment for test automation are:

- This is a dedicated environment for the testing and owned by the testing team.
- Test data from manual functional and regression testing can be used for automated testing.
- Support from the wider testing team is available for automation, for example functionality walkthrough and test data creation.
- Stable SUT avoids frequent changes to automation scripts.

The challenges of using the test environment for test automation are:

- The latest source code from the development environment may not be available.
- Test environments are not a true representation of production.
- Testing tool installation and configuration requires full access to the environment.
- Regular backup for the automation test scripts may not be possible.

6.1.3 Production support or pre-production

The production support or pre-production environment is generally a 'live-like' or sometimes a scaled-down version of the production environment. The purpose of this environment is to support the 'live' system. This is generally used as the final gate to deliver to the production environment. This environment is used for UAT in many places and supports the investigation of issues raised within the production environment. This is also used as an isolated platform for upgrades, patching, bespoke code and data releases, which are to be informally applied before being applied to production. This environment can also be used for usability and accessibility testing if required. This environment is a better representation of the target production environment and can be used to support the key non-functional test stages.

The benefits of using the pre-production environment for test automation are as follows:

- Pre-production and production environments are identical and are a true representation of production for automation.
- The pre-production environment helps to run test automation in an exactly production-like environment.
- The latest features are available for test automation, despite being not yet available in the production environment.
- Pre-production environment is widely used for automated performance testing.

The challenges of using the pre-production environment for test automation are the following:

- The latest source code from the development and test environments are not available for test automation.
- The pre-production environment is shared with other teams such as performance testing and UAT, and this can impact the stability of the environment.

6.1.4 Production

This environment will be used to deploy the final solution and may be used to support some of the non-functional test stages, such as penetration testing and security testing.

The benefits of using the production environment for test automation are as follows:

- Live features are available in the production environment.
- The SUT is already available to the end-users and customers, and so it is a true representation of the final solution.
- The production environment is widely used for product testing, for example beta trialling.
- Automated regression testing on the production environment provides a quick view of the SUTs.
- The production environment is widely used for automated security testing and disaster recovery testing.

The challenges of using the production environment for test automation are the following:

- Any testing in the production environment can impact the stability of the environment, and it can affect the end-users and the business.

- The solution is already live, and any defects identified in production take time to fix and roll-out.

6.2 SUMMARY

This chapter is an introduction to different environments and their roles in the success of test automation. It addressed some of the benefits and challenges of using different environments for successful test automation. Next, we will explore candidates for test automation.

7 CANDIDATES FOR AUTOMATION

Tests and test cases are the first candidates for test automation; however, they are not alone. Tasks such as test data creation, requirement traceability and test reporting are also great candidates for automation. This chapter describes various steps in choosing the right items for test automation and what you should avoid. Some test cases are suitable for manual testing, and some are unsuitable for automated testing as they do not provide enough benefits. This chapter helps test managers and Scrum masters to understand the factors to consider when choosing suitable candidates for test automation.

Over the implementation period, the team identify more candidates for test automation. The scope of automated testing grows from functional and regression testing to additional areas such as requirement traceability matrix, test data creation or non-functional testing. Figures 7.1 and 7.2 show how test automation grows over a period, and additional suitable candidates add to automated testing.

Figure 7.1 Test automation: initial days

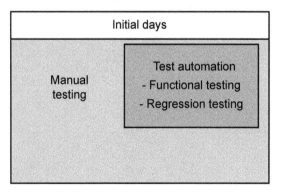

7.1 WHAT SHOULD BE AUTOMATED?

In the previous chapters, we have addressed the benefits of test automation, using CBA, ROI and so on. It is essential to know how to choose the suitable candidates for automation. The scope and benefits of test automation are beyond functional and regression testing. More areas can be considered for test automation and automated

Figure 7.2 Test automation growth

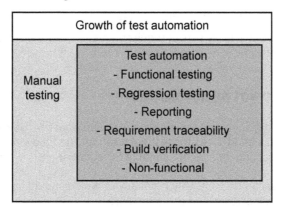

testing, such as smoke testing, build validation and test data creation. Here are some candidates that can be considered for test automation:

- Business features or user flows that cause considerable damage to the business if they fail; automated testing helps to validate these features frequently.

- Tests that need to run against every build or release of the application, such as smoke tests, sanity tests and regression tests.

- Tests that need to run against multiple configurations, such as different platform and browser combinations.

- Tests that involve large sets of test data or inputting large volumes of data, for example filling in very long forms, as test automation helps to achieve extended coverage, reduction of effort and better reliability.

- Reporting requirements, as automated test reporting reduces manual intervention and generates frequent and up-to-date reports.

- Functionalities and test cases that provide immediate benefits, for example smoke and regression tests in DevOps to examine if the deployed build is stable.

- Tests that require overnight dedication, such as batch processes and reports.

- Repetitive actions such as creating test data for every iteration.

- Repetitive tests such as common tests or prerequisites for multiple tests.

- Tests executed with different groups of data, multiple browsers, environments, complex business logic, calculations, different sets of users, special data, on compliance-related matters such as World Wide Web Consortium (W3C) standards, and security testing.

- High-risk test cases and test cases based on high human error hazard potential.

- Extensive tests with a large set of test data and various permutations and combinations.

- Non-functional testing, such as performance testing, load testing, soak testing and stress testing.

- Tests that are time-consuming and hard to do manually.

- Traceability matrix of large sets of requirements as updating the traceability matrix after every test run is cumbersome.

7.2 WHAT SHOULD NOT BE AUTOMATED?

Test automation generally provides high returns and benefits. However, not all tests and SUTs are suitable candidates for test automation. The following factors should be considered when deciding what is not suitable for automation:

- Test cases that can be done only manually or which should be done manually, for example user experience and usability tests.

- Low value and low priority tests as they do not provide sufficient returns.

- Tests that run only once; the exception is data-driven tests with large sets of inputs.

- Tests involving high additional licensing costs such as special reports.

- Tests that require frequent user intervention.

- Tests requiring ad hoc or random testing based on the domain knowledge or expertise, such as exploratory testing.

- Frequently changing features and outcomes, as they are difficult to automate and validate the expected result against the actual result.

- Tests without predictable results such as scientific simulation.

- Test results requiring instant visual or manual confirmation, for example colours.

- Tests that need a high level of effort to automate and low return such as user interface testing.

Table 7.1 is a sample template to identify the suitable candidates for test automation. This template also helps to eliminate items that are unsuitable for automation. Test leaders and automation test managers need to customise this template with the right parameters, subject to their products and solutions.

7.3 SUMMARY

Choosing the right candidates for automation is critical for the success of test automation. This chapter provided factors you should consider when choosing and eliminating candidates for automation. The chapter also provided a template to perform the selection of suitable candidates systematically.

In the next chapter, we will explore achieving test coverage through test automation.

Table 7.1 Candidates for automated testing

ID	Description	Complex business logics	Multiple users	Multiple browsers	Different data set	Test data volume	Special data	Multiple environments	Dependency	Candidate for automation	Comments
1	Des 1	Y	Y	Y	Y	Y	Y	Y	Y	Y	
2	Des 2	Y	Y	Y	Y	N	N	N	N	Y	
3	Des 3	N	N	N	Y	Y	Y	Y	Y	Y	
4	Des 4	N	N	N	N	N	N	N	N	N	
5	Des 5	N	N	N	N	N	Y	N	N	N	
6	Des 6	Y	Y	Y	Y	Y	Y	Y	Y	Y	

8 TEST AUTOMATION AND TEST COVERAGE

Test coverage is used as an indicator to measure the quality and effectiveness of software testing. It is a method used to determine coverage of tests of the application code and amount of code that is exercised when the tests are run for example. If there are 100 requirements and 100 tests created for 100 requirements with 90 tests being executed, then the test coverage is 90 per cent. This helps to understand the gap in testing and create additional tests for solution coverage. Test coverage helps quality assurance ensure quality of the test, gaps in test cases, code coverage, testing scope measurement and defect prevention.

Test coverage is widely used for test completeness and test automation is a widely accepted method to increase test coverage quickly. Test coverage and test metrics provide enormous confidence to the stakeholders. Test automation provides 100 per cent test coverage in some areas such as unit testing, performance testing, security testing and data validation. The most commonly used metrics are related to test execution efficiency, automation coverage, effort reduction, additional test cycles achievement, defect detection trends (DDT) and so on. Chapter 13 addresses test automation metrics in detail. This chapter addresses the value of test automation in test coverage.

8.1 TEST COVERAGE

One of the widely accepted ways to measure test coverage is based on functional and non-functional requirements. The functional and non-functional requirements drive what is expected from the application. Functional requirements describe the application's expected behaviour, and non-functional requirements describe the reliability and sustainability of the application.

Requirement coverage is not always achieved through testing or test automation, as many requirements such as legal, standards and compliance are achieved through static testing and verification. A test traceability matrix is the most common method to track the requirements until the final application, and test automation ensures test coverage is met quickly and consistently. Various tools such as Micro Focus ALM, Microsoft Azure DevOps and Jira are available for the requirement traceability matrix (RTM).

'Traceability is the degree to which a relationship can be established between two or more work products. A traceability matrix is a two-dimensional table, which correlates two entities, for example requirements and test cases. The table allows tracing back and forth the links of one entity to the other, thus enabling the determination of coverage achieved and the assessment of impact of proposed changes' (according to ISTQB Glossary, https://glossary.istqb.org/app/en/search/).

Requirement coverage and code coverage are two different concepts, even though they are often used to measure the same result. Code coverage is a common unit testing practice that is widely used to measure how much of the code is actually being used in the unit tests, how much code is not being used in unit tests, and also shows which branches are and are not being covered. Unit testing is part of the software development activity, and the responsibility lies with the programmers or developers. There are various tools available to check and optimise the unit code coverage.

Requirement coverage is the responsibility of the testing team. They validate that the tests have covered all the test conditions that were derived from the requirements and other acceptance criteria. A requirement traceability matrix is often used to trace the requirements until acceptance and is a key parameter in deciding when to stop testing.

This section will address some of the most common test coverage methods widely used in the industry and added value provided through test automation.

8.1.1 Unit test coverage

Unit tests are typically written by programmers to validate the correctness of the code. This is to verify that the code has been programmed in line with the program standards and LLD or detailed design document (DDD) derived from the product backlog or the requirement specifications.

LLD or DDD defines the design of the physical architecture of the selected logical option, server types, network architecture, storage requirements, interfaces, network design and so on.

Unit testing is generally expected to be 100 per cent automated by using appropriate tools. However, if the effort versus value benefit does not weigh up, unit tests should not be conducted, for example a simple call to API with no business logic in the code will not be unit tested as it would be a significant effort for very little benefit. Code coverage basically shows how much of the code is being used in the unit tests. Code coverage also demonstrates which branches in conditional logic are not being covered. A line coverage measures how many statements or lines of code are covered. Branch coverage ensures that each one of the possible branches from each decision point is executed at least once.

CIT, also known as unit integration testing, is to verify that the program components are all present and can work together. The CIT should be 100 per cent automated, where possible. Product-based organisations place high emphasis on unit testing. Unit tests are generally either performed manually or automated, with a wide-ranging set of tools available for automation. Some organisations pay less attention to unit testing as it is time-consuming and slows down the solution development progress. Unit testing improves the quality of the software solution and identifies the defect much earlier in the defect cycle. Unit tests are often developed at the LLD and performed at the micro-level. The key inputs for the unit test are the product backlog catalogue or requirements, style guide, code or component units and LLD. They are generally performed on the development environment and development workstations. Off-the-shelf software tools and package vendors generally provide an inbuilt testing framework known as the automated test framework (ATF) for unit testing.

> The purpose of unit testing is to detect implementation errors in units of code, for example a method or stored procedure.

There are many industry-wide tools and frameworks available for unit testing, for example Java-based and .Net based, to support developers. Many packaged tool providers such as ServiceNow and Oracle provide their own unit test suite to support the unit testing. These test suites, generally known as ATF, are inbuilt in the solution and provided as part of the packaged tool or COTS software.

The key purposes and benefits of unit testing are:

- Good practice, providing quality at a micro-level.
- Finds the defect earlier and at the design level.
- Each basic component or bespoke unit of code functions according to its design specification.
- Exposes issues in the interactions between component units.
- Validates component performance.
- Accessibility compliance for custom code.
- Validates coding standards.
- Checks automated build passed.
- All installation, configuration and data transformation processes are successfully performed.

Automated unit testing reduces manual intervention, and the tests are coded in a programming language with tools and execution against the code base as part of the build process. These automated tests are executed every time after a code build or release, and results are reviewed for errors. Developers use the unit test result to validate the quality of their code.

Automated unit testing provides huge benefits to the test coverage. The key benefits are:

- Used frequently without additional costs compared to manual testing
- Makes the development process more flexible
- Automatically checks the standards, compliance and guidelines
- Makes it quicker to find problems in the development cycle
- Uncovers errors during coding and speeds up development
- Tracks and monitors code quality all the time

- Quicker build validation, easy refactoring and cleaning up existing code
- Quick and detailed test coverage supporting quicker bug fix, hotfix and release

8.1.2 Functional and regression test coverage

Functional testing provides test coverage on functional requirements. The test scenarios and test cases are mapped to requirements to ensure that they have coverage on specifications. Automated test scripts are directly mapped to functional test cases and requirements providing a good indication of test coverage. A well-designed automated framework helps to achieve test coverage quickly and efficiently. An automated requirement traceability matrix provides real-time coverage on requirements and is very useful for large projects and product development.

Test cases are prioritised, and a subset is selected for regression testing. The scope of the regression test pack varies subject to the scope of the release, such as full release, hotfixes, patch release and regular releases. An automated regression pack derived from requirement mapped to the regression pack provides immediate visibility on test coverage. Manual regression testing is monotonous, and less effective once the frequency of the test execution is high, if done incorrectly. Testers tend to ignore the test, and test execution becomes a 'routine' rather than a thought-through activity. Test automation is a good method to make it more effective and extensive using different data sets. Test automation is the most suitable test coverage method for regression testing.

In regression testing, the previously executed test cases are re-run, time after time, on a new build, which is time-consuming. The test cases and expected results are available from the previous execution, and automating these test cases is a time-saving and efficient regression test method. Selecting test cases for automation is performed based on various factors such as risk, ROI, the time factor, dependencies and high volume of data.

Smoke testing is generally used to validate whether or not the deployed build is stable and ready for further detailed testing. It is generally performed by the build and deployment team or the test team. Most of the smoke tests are automated, and they are expected to provide quick results. In general, smoke tests are 100 per cent automated and provide full test coverage on build validation.

Automated functional tests provide quick results and full test coverage. Integration testing is an extension of functional and system testing, and the same automation principles are applicable for integration testing. Technical integration tests such as web services testing, API testing and interface testing are generally fully automated and provide 100 per cent test coverage.

The key features of automated functional testing in order to provide test coverage are:

- Provides consistent test coverage
- Automatic coverage of the standards, compliance and guidelines
- Complete test coverage on smoke testing
- Quick and detailed test coverage of functionality

8.1.3 Non-functional testing coverage

Non-functional testing is checking the non-functional aspects of the system, such as performance, usability, reliability, security, accessibility and infrastructure of a software application. It is to test the readiness of a system according to the non-functional requirements. Test automation is a suitable solution for a wide range of non-functional testing, as manual testing is not effective or feasible. Automated non-functional testing is quick and provides full test coverage on requirements. Most of the non-functional testing is fully automated and provides good test coverage, for example performance testing, security testing, browser testing and mobile testing can be fully automated and provide 100 per cent test coverage. However, usability testing, user experience and infrastructure testing are difficult to automate and provide less test coverage.

Key benefits of automated non-functional testing to test coverage are:

- Automated performance, volume and scalability testing provide high test coverage compared to manual testing.
- Tools and automation are essential for consistent test coverage in security testing, for example penetration testing or vulnerability scan.

8.1.4 Challenges in measuring test coverage

Test automation brings speed, confidence, efficiency, reliability and completeness to test coverage and is essential in software testing. However, there are a few challenges in using automation for test coverage.

- When requirements are incomplete, testing is difficult to automate and is unable to provide accurate test coverage.
- The low priority requirements may not qualify for automated testing. This can reduce the level of test coverage through test automation.
- Test coverage in different test stages has a different meaning. Test automation is high for automated unit testing. UAT is generally performed manually and may not be a good candidate for automated test coverage.
- The requirements that are hard to automate are not considered for automation and automated test coverage.
- One hundred per cent automation and test coverage through automated testing are difficult to achieve, for example in usability testing.

8.2 TEST AUTOMATION METRICS

Test automation metrics are vital for test coverage and in ensuring the quality of automated testing. Test automation is a huge investment, and it requires continuous monitoring, tracking and measuring to ensure that it returns the expected benefits and provides ROI. The investment in test automation and automated testing cannot be justified if it cannot be measured. Test metrics are described in more detail in Chapter 13.

'You can't manage what you can't measure.' (often attributed to W. Edwards Deming, the statistician and quality-control expert credited with having launched the Total Quality Management movement, and at other times it is attributed to Peter Drucker).

The investment in test automation and automated testing cannot be justified if it cannot be measured. Test metrics are further described in Chapter 13.

- Percentage of test automation coverage
- Percentage of tests passed versus failed
- Percentage of automated tests passed versus failed versus manual tests passed versus failed
- Number of defects found by automated scripts over time to assess the effectiveness of the script
- Number of defects found in manual testing versus automation
- Requirements coverage by automation
- Test automation effort versus manual testing effort

8.3 SUMMARY

In this chapter, we addressed how test automation enhances test coverage. The investment in test automation cannot be justified if you are unable to measure it. Test automation metrics provide a reliable method to measure the return on any investment, understand which parts of the testing are working, and support improvement.

This concludes Part One of the book, which had the objective of assisting the test practitioner in understanding test automation and automated testing for implementation in their organisation. Part Two of this book address practical aspects of test automation, that is, the 'How' and 'When' of test automation.

PART TWO
HOW AND WHEN TO DO AUTOMATION

9 TEST AUTOMATION SUBJECT MATTER EXPERT

In Part One of this book, we explored the strategic and leadership view of test automation: what is test automation, why is it required and why is test automation different from automated testing? The chapters in Part One support the decision-making process, and once the decision is taken that test automation is the suitable solution for the organisation, the next step is to explore the details of how to approach it.

Part Two addresses how to approach test automation and when to perform the activities. This part of the book helps test managers and Scrum masters in building successful test automation for their project or product and the organisation, explaining how to measure the success in test automation with a set of metrics and key performance indicators (KPIs).

The people who build automated testing are widely known as test automation SMEs, automation testers, technical testers or test automation engineers. This book uses 'test automation engineer' to ensure consistency. This chapter provides insight on how to build and lead a successful career in test automation and the skills required in order to enhance career progress, as well as lateral career movement from coding or manual testing to automated testing. The focus also remains on the competency matrix for a test automation engineer.

> A subject matter expert has a sound knowledge of and expertise in a specific area, and they are considered an authority in this area. A test automation SME is an expert in test automation, the process and technology.

9.1 ESSENTIAL SKILLS FOR A TEST AUTOMATION ENGINEER

The test automation engineer prepares, performs and reports on the specialist testing of products and services by utilising an appropriate tool or technical skill in close liaison with the project team. An experienced test automation engineer in their field is capable of managing, designing, developing, executing and reporting on automated testing utilising the tools or scripting skills.

Test automation engineer is the 'person who is responsible for the design, implementation, and maintenance of a test automation architecture as well as the technical evolution of the resulting test automation solution' according to ISTQB Glossary (https://glossary.istqb.org/app/en/search/).

Developing code and script is a developer's work in any software solution, and manual testers validate the developer's work based on the requirements. However, a test automation engineer performs both a developer's job and a manual tester's job by validating the requirements manually and using an automated script developed for testing. Writing code and debugging is part of a test automation engineer's job, and they require software programming or coding knowledge to perform this well.

Test automation engineers are often asked to do manual testing. However, 'one size fits all' is not a good approach as both require different skills. Manual testers may not have the right test automation skill, and test automation engineers often get bored doing manual testing in the long term. The list below and Figure 9.1 present the essential skills required for a test automation engineer:

- Testing skills
- Requirements knowledge
- Manual testing
- IT skills
- Knowledge of the SUT
- Problem-solving skills
- Expertise in test automation tools
- Knowledge of software development processes and approaches
- Documentation skills
- Knowledge of the latest test automation trends
- Reporting skills
- Decision-making ability

The following sections describe the essential skills required for a test automation engineer in detail.

9.1.1 Testing skills

A tester verifies the quality of a test object through the preparation and execution of appropriate tests. A test object can vary from a feature, function or interface between two applications, to a total end-to-end solution. The test object can be functional features or non-functional aspects such as backup, restoration, performance, accessibility or

Figure 9.1 Test automation engineer essential skills

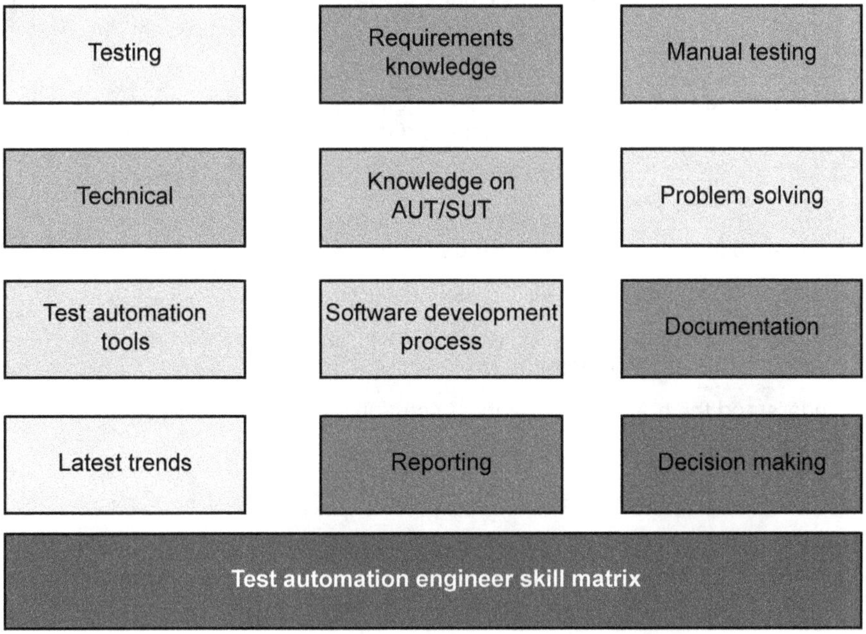

Testing	Requirements knowledge	Manual testing
Technical	Knowledge on AUT/SUT	Problem solving
Test automation tools	Software development process	Documentation
Latest trends	Reporting	Decision making

Test automation engineer skill matrix

compatibility. In addition to testing skills, a test automation engineer is expected to have good programming or coding skills.

The following testing skills and knowledge areas are prerequisites for a good test automation engineer:

- Ability to understand project architecture such as high-level design, low-level design and network architecture
- Experience in projects, the test life cycle and the test process
- Knowledge of release, change and configuration management
- Expertise in development approaches such as Waterfall, Agile, DevOps and CD
- Knowledge of management information reporting, communication and stakeholder management
- Ability to develop automation test strategy, plan and approach
- Functional, non-functional and manual testing expertise
- Estimation, documentation and reporting skills
- Experience in defect reporting and tracking

The skills that are expected from any test automation engineer are listed below. Building expertise in these areas will help in maintaining a successful career in test automation.

Technical skills required for a test automation engineer:

- Expert in the use of a specialist tool, for example Selenium, SOAP-UI, Eggplant, Tosca for functional test automation, Micro Focus LoadRunner or JMeter for performance testing, and JAWS for accessibility testing.

- Analyse functional or non-functional requirements and, where necessary, expand on these in liaison with stakeholders and the business analyst.

- Create automated test scripts and scenarios as needed to test functional or non-functional requirements.

- Execute technical tests, analyse and present results.

- Liaise with SMEs to diagnose issues uncovered by testing.

- Install technical testing tools, carry out the proof of concept work for tool use and configure the test environment.

- Understand the main components of solution or system architecture.

- Able to identify where the technical testing fits within the SDLC.

- Experienced in coding or scripting language.

- Produce a technical test plan (e.g. for automation or performance testing) and technical test exit report.

- Create automation test frameworks and coach their team on the use of the framework.

- Able to approach technical testing challenges both methodically and creatively.

- Act as a mentor or coach to less experienced test automation engineers.

- Report on progress and escalate issues.

- Experience in at least one method of creating and loading large volumes of test data.

- Undertake peer reviews.

- Apply Agile techniques such as leading in daily stand-ups and scrums, and use of backlogs.

- Hold the ISTQB Advanced Test Automation Engineer or ISTQB Advanced Technical Test Analyst and tool-specific certifications.

9.1.2 Business domain and requirement knowledge

Business domain and requirements are classified as either functional or non-functional requirements, and it is essential that a good test automation engineer understands them and can assess their impact on testing and test automation. Functional testing verifies application behaviour, and non-functional testing verifies the sustainability of the system, such as performance, security and scalability. Engineers also require domain knowledge (such as in informatics, retail or banking) to design effective TASs.

A scientific software product company is a leading market player in delivering products and solutions for research (simulation) and education. Their products have been developed over two decades of research and development. These products were mostly developed by university students for research purposes and were enhanced further into fully operational products. Testing these products was very challenging as there were no definite outcomes and required deep domain knowledge. The expected results varied from time to time. Most of the testing was performed manually by scientists or subject matter experts along with core testing professionals. Testing was time-consuming and heavily dependent on domain knowledge.

As part of test automation, the scientists and SMEs received training on the testing tools. Their in-depth knowledge of functional requirements and scientific domains helped them to automate the testing. The test organisation managed to automate all manual testing using the software and skills of the SMEs.

An experienced test automation engineer understands requirements and their impact on automated testing. This helps them to prioritise and automate critical business features. Knowledge of business requirements helps the test automation engineer to:

- Identify business-critical scenarios for test automation.
- Comprehend how the solution works to develop effective automated tests.
- Identify the right test automation framework.
- Understand business priorities.
- Design scalable automation suites.

9.1.3 Manual testing skills

Manual testing is the prerequisite and essential skill for any testing – functional, non-functional or automation. The test automation engineer has a thorough knowledge of manual testing to create effective automated testing solutions. Manual testing helps them to understand the domain and features better. Many teams prefer test automation engineers to test the solution manually before creating automated testing. The manual testing process and steps assist the test automation engineers to understand how the solution works.

Manual testing skills are key factors in the decision-making process for what needs to be automated. Manual testing skills support test automation in the following activities:

- Identifying candidates for manual and automated testing
- Identifying scenarios for regression testing
- Choosing candidates for high ROI
- Separating the time-consuming tests such as data-driven or multi-browser

- Better understanding of manual and automated testing
- Manual intervention to interpret automated test results

9.1.4 Technical skills

The de facto skills for automated testing are technical and coding knowledge, both of which are difficult to learn and excel at. A tester with good manual and technical skills succeeds in automation, as this is one of the primary reasons why many manual testers with expertise in coding make a lateral career move to automated testing. A good starting point in automated testing is being involved in maintaining and enhancing the existing automated test suite. This builds expertise in how scripts work and enhances them to be more effective. The key IT skills required for automated testing are:

- Proficiency in programming languages such as Java, .NET, Python, Perl, C/C++, SQL, XML and HTML
- SQL to retrieve specific information from databases
- Good coding skill
- Good understanding of automated test scripts
- Expertise in test automation frameworks
- Detailed knowledge of automation tools
- Debugging skills
- Knowledge of platforms such as on-premises, cloud and hybrid
- Experience in test environments such as development, test and production
- Knowledge of IT infrastructure
- Experience in tool installation and configuration
- Ability to read and understand SUT architecture
- Experience in automation framework architecture
- Expertise in PoT and PoC

A government organisation spent billions annually in meeting its communication objectives, the bulk of which was in print communication. This leading and reputed organisation analysed spending behaviour and ROI and decided to use IT to reduce costs. One major outcome was the idea to develop a portal, attract the public and reduce the print media. Testing was a core activity for the IT solution to ensure the solution met expected results accurately. Test automation was heavily used to reduce the cost of testing in the long run.

IT knowledge is a key skill for successful automated testing. The automation team used their prior experience in development and coding to support both the development and testing teams. As part of defect reporting, the automation team managed to provide root-cause analysis to the development team. Technical skills helped the automation team create a competent automation framework.

9.1.5 Knowledge of system under test

Test automation engineers need to acquire good knowledge of SUT, application behaviour, technology, architecture and business requirements to develop effective automation solutions. The following areas of the SUT are particularly relevant:

- Functional and non-functional requirements
- Programming languages used to develop the SUT
- SUT architecture
- Interfaces, web services and APIs used
- Functions and features of the application
- Database
- Current testing process of the SUT
- Release environments
- Complexity of the application
- Domain and sector
- Regulations such as Web Content Accessibility Guidelines (WCAG) and Health Insurance Portability and Accountability Act (HIPAA)
- Compliance, such as for browsers, operating systems, mobile and web

These areas can influence and impact the design, development and execution of the test automation suite.

9.1.6 Problem-solving and decision-making ability

Problem solving is a key skill for software coding and is essential for any test automation engineer. Problem solving involves:

- Problem definition
- Investigating the cause of the problem
- Identifying solutions
- Prioritising solutions
- Selecting a solution
- Implementing the selected solution

It requires:

- Active listening and analysis
- Research and investigation
- Creativity

- A positive attitude and aptitude
- Communication

Decision making is an important part of coding, test management and system design. This is an essential skill expected from all levels of test automation professionals. Decision making is performed algorithmically or heuristically. An algorithm is a precise set of rules and conditions that never change, while a heuristic is a set of rules that may change over time as conditions change. It is easy to think of automating tasks traditionally performed by people, but wrong decision making placed into a machine can become a critical quality issue of the system.

Problem-solving skills enable handling of unexpected situations or difficult challenges. Problem solving is a key check involved in the hiring process of test automation engineers; at junior levels, the focus is on aptitude and technical abilities, and at senior levels it is about decision making. A few hiring questions related to problem solving and decision making are shown below.

Junior levels:

- What is the 5-digit maximum prime number? How did you arrive at the answer?
- What is the minimum number of satellites required to cover the whole earth, and why? Explain your rationale.

Senior levels:

- Give an example of when you ran into a problem on a project and how you solved it.
- How would you manage a dissatisfied manager, stakeholder or customer?

Test automation engineers need good problem-solving and decision-making ability and technical knowledge to be successful. In the automation test design phases, test automation engineers come across scenarios to create scripts that cover many possible outcomes, including positive, alternate and error handling scenarios. Good problem-solving skills are an advantage in developing better and more effective automation test scripts to handle real-time scenarios.

9.1.7 Knowledge of test automation tools

Test automation engineers are expected to be proficient in the use of automation tools and underlying programming languages. In practice, most projects and organisations choose widely available tools and frameworks in the industry, and test automation engineers are expected to be familiar with them.

Most job descriptions clearly demand expertise in these testing tools as they have active community support and provide immediate results. The bespoke (customised based on client requirements) test automation tools and frameworks provide numerous benefits and advantages. Expertise in these tools reduces automation effort and improves the ability to conduct a PoC. Prior knowledge of the automation tools avoids a PoC for known platforms and provides immediate returns.

In the current environment, companies and stakeholders expect quick ROI and speedy and high-quality deliverables, and expertise in automation tools is an added advantage to meeting these stakeholder expectations. Being a skilled test automation engineer with in-depth technical knowledge and experience in leading/popular automated testing tools is mandatory. test automation engineers are also required to advise on the tool setup and infrastructure requirements such as hardware, licensing, installation and configuration.

9.1.8 Knowledge of software development process and approaches

Software development approaches and their impact on test automation is covered in Part One of this book. The flavours of Waterfall and Agile are widely used in the industry, and test automation engineers should demonstrate good knowledge of software development approaches like these. The project and the organisation expect the automation team to be well versed in popular development approaches and deliver in line with them.

Waterfall, DevOps and Agile work differently and a lack of experience in these processes will impact the ability to deliver automated testing. Good knowledge of the development process enhances a test automation engineer's career, and certifications such as Certified Scrum Master or Certified Product Owner provide an additional advantage.

Test automation engineers are also expected to have expertise in the different phases of testing such as test planning, design, execution, defect management and reporting. Automated testing follows a similar life cycle to software development: planning, design, execution, maintenance and closure.

9.1.9 Documentation, communication and reporting skill

Documentation and reporting are two key skills for any test automation engineer. Automation test strategy, plan and approach are the essential documents involved in automated testing. Good documentation communicates a clear message to the stakeholders. Automation funding and approval are often subject to good, clear documentation, and poor documents impact the decision-making process.

Another key document is test reporting, most of which can be automated. Automated test execution reports are easy to manage as they are scheduled to be created and circulated automatically without any human intervention.

Documentation is largely ignored by many development and testing projects. However, well-documented automated tests add value in the decision making for PoC, maintenance and bug fixes. Test automation engineers are the developers of test scripts, and their code must be documented and commented in peer reviews to support team collaboration.

Good communication and collaboration are essential skills for a test automation engineer. This is important for team bonding among test automation engineers and other team members. Test reports with effective communication are vital in establishing trust with stakeholders. Refer to stakeholder management in Chapter 1.

9.1.10 Knowledge of the latest trends in test automation

Automated testing is continually evolving, from the 'record and playback' to advanced automated suites, simple tools to highly sophisticated frameworks, and manual testing to codeless scripting. The vendors look beyond the current market and release automation tools for future needs. A good and successful test automation engineer stays ahead of the trends and updates themselves in order to be on top of the latest developments. The best practices, standards, guidelines, automated testing frameworks, tools and script development processes are constantly evolving. Test automation is moving swiftly to AI, ML and robotics, and it is important to be aware of the latest trends, tools, products and so on. The latest trends provide the opportunity to understand the evolving challenges and act accordingly.

Technology changes rapidly and is challenging to keep up with. Professional bodies such as Agile Alliance, Association of Computer Engineers and Technicians, and BCS organise events to create awareness and discussion on emerging trends. In addition, books, points of view and articles from reputed practitioners are good sources of information. Well-known publications and online websites such as LinkedIn publish the latest trends on a regular basis. Tech communities are a good source of the latest trends along with IT conferences. Tool vendors actively reach out to the automation test practitioners to evaluate new tools, giving them a wonderful opportunity to experience the upcoming tool features hands-on.

Functional testing tools over the last two decades have evolved, for example WinRunner as the predominant tool to QTP and thereafter Selenium WebDriver, which is in high demand of late.

9.2 BUILDING A SUCCESSFUL CAREER IN TEST AUTOMATION

Building a successful career in automated testing and becoming an SME in test automation is a disciplined and structured activity.

'Building a successful career in test automation is like a Marathon and not a Sprint' (attributed to Peter Hanson, Programme Test Manager).

This section provides some guidelines and structure to build the essential skills discussed earlier in this chapter in the achievement of this goal.

1. The fundamentals of testing. Building excellent knowledge of the fundamentals of testing and test automation is essential:

   - Test certification, for example ISTQB. Certifications help build fundamental and advanced knowledge systematically. It is often considered a prerequisite to test automation jobs and the foundation for building a successful career.

- Websites related to test automation – There are plenty of automation-specific websites to learn from.

- Books – Another excellent source of knowledge.

- Attending seminars – This helps to build knowledge on various products and services in test automation. Testing conferences are a good place to network with fellow test automation experts and discover the latest trends.

- Test strategies and plans – Planning is key to any test automation, and test automation engineers and SMEs are expected to have good experience in test automation planning.

2. Fundamentals of test automation and automated testing. The following areas are critical for test automation:

- Attaining automation certification

- Learning programming languages

- Building coding skills

- Understanding test automation frameworks

- Static review and code review of test scripts

- Modifying existing scripts

- Using a sandbox area to practise automation

3. Developing a framework (see Chapter 12):

- Seeking out a mentor for guidance

- Finding an application to automate, including login

- Creating a plan and reviewing with your mentor

- Creating a simple framework

- Developing functions such as exception handling and reporting

- Learning tools such as Selenium, Appium, Eggplant and Tosca

4. Job hunting. This can also be used to test the market and test your value in the market:

- Create a CV.

- Get a reference or two.

- Maintain an active LinkedIn profile.

- Register on job sites.

- Attend interviews and attend mock interviews.

- Join the interview panel at your organisation; you can always learn from good candidates.

5. Growing in your career:

- Learn new frameworks.

- Learn new tools.
- Join communities.
- Create visibility through networking.

9.3 SUMMARY

This chapter summarised the essential skill sets that are expected or required for a test automation engineer. This chapter also provided some tips on building a successful career in test automation.

The next chapter addresses widely used test automation tools available in the market. To be a successful test automation engineer, one needs to have a crystal-clear understanding of the 'whats', 'hows' and 'whens' of test automation, along with sound technical expertise and tool experience.

10 TEST AUTOMATION TOOLS

In Part One, we explored the testing tool selection process. This chapter addresses different cost parameters to be aware of before procuring test automation tools for your organisation and briefly covers some of the widely used tools in functional testing, non-functional testing and test management. The demand for new automated testing tools keeps on changing with technology evolution, development approaches and market changes. The tool vendors continually deliver new tools to meet the market demand, and as such, it is difficult to list down and maintain a set of the latest tools at any point in time. The second part of this chapter lists a set of widely used tools for you to familiarise yourself with. This is not a recommendation of any tools over others or an endorsement of tool features.

Many organisations procure test automation tools without proper research, not understanding the licence model, professional support required or availability of in-house technical expertise and, as a result, end up disappointed. The tools themselves are often blamed after money and effort have been spent. Tool vendors often use terms such as 'seat licence', 'concurrent licence', 'shared licence', 'annual licence' or 'professional service', with costs associated with each. It is confusing and difficult to understand and decide on what is required and fit for purpose within an optimised cost. Many open-source tools have hidden costs such as add-ons or third-party solutions for features such as reporting and analysis.

The procurement process can be painful and time-consuming when the test automation manager drives it. Many organisations have a procurement process and team to assist the test automation team in gathering the information and procuring the tool on their behalf. Setting up and approving a new tool vendor takes time and a lot of follow-up, especially in big organisations. Also in many cases, the licence renewal process starts months before the licences expire.

The first part of this chapter describes different licence models, one of the major cost factors in most of the test automation activity. The second part introduces popular and common test automation tools available in the market. The third part of the chapter explains different cost factors to be considered for overall test automation tool costs.

10.1 UNDERSTANDING LICENCE MODELS

This section is intended to explain various licensing models and costs involved in procuring tools.

Firstly, some common terminologies related to tool licence management may be useful:

Software licence manager. This is a management tool used to control where and how software products are used. Licence managers protect, comply, monitor and track software licence agreements and usages. Licence managers offer a wide range of licensing models, such as product activation, trial licence, subscription licence, feature-based licence and floating licence. In many cases, the licence manager checks if the licence is in place remotely through the internet and tracks usage of the software. This tool is generally created by a vendor for their product or a bundle of products.

Software asset management or log. This is used by end-users to manage the software they have procured or licensed from many different software vendors. These are used to reconcile software licences and are largely managed on a spreadsheet for most small organisations. Large and medium size organisations maintain a 'configuration management database' or 'CMDB repository' that acts as a data warehouse for asset management. The CMDB stores information about the IT environment, the components, hardware assets, software assets (referred to as configuration items) and the licences.

Product activation. This is a licence validation mechanism, and it prevents the violation of the software licence. Activation allows the software to stop blocking features from the user when they move from using its free version to purchasing a licence.

Product key. This checks and confirms that the software is original from the vendor. For example, a physical copy of Microsoft Windows comes with a 25-character product key on a label or card inside the box, or the performance testing tool LoadRunner licence key can be downloaded online for an existing customer.

The list below covers different types of licence models and associated costs, often used in test automation tools:

- **Public domain.** These tools are available in the public domain and there is no ownership such as copyright, trademark or patents used to modify and distribute them. They are free to use.

- **Licence-free.** These tools are intended for free use, modification and distribution as long as the copyright is retained by the original owner.

- **Free and open-source software (FOSS).** These tools are freely licensed to use, and the source code is encouraged to be shared voluntarily with the community to improve the software. A few examples of FOSS are: Selenium, Appium, Apache JMeter, SoapUI and TestNG.

- **Original equipment manufacturer (OEM).** An OEM licence is generally installed by the manufacturer on new devices that are not usable on other machines, for example Microsoft Windows and anti-virus software are pre-installed on new Windows laptops. This licence is generally available for a short period unless renewed. Another example is the many browsers such as Chrome, Edge, Firefox and Safari that are pre-installed with compatibility and accessibility testing tools.

- **General Public Licence (GPL).** This is a free licence for end-users to learn, share and modify the software.

- **Multi-licence.** These distribute software under two or more different sets of terms and conditions, for example the software vendor distributes the same software to two different sets of end-users as GPL and OEM. The features of the software are the same. However, the usage is different.

- **Permissive software licence.** A free software licence with minimal restrictions of usage, modification and distribution.

- **Proprietary software.** This is under strict copyright licensing, and the source code is usually hidden from the end-users preventing modification, for example, Tosca and Eggplant.

- **Subscription licence.** This is a periodic renewable licence, for example users of Azure DevOps can either pay month-to-month through Azure or buy classic software licences, which require a 3-year commitment. BlazeMeter is another tool based on an annual subscription.

- **Concurrent/floating licence.** A software licence that is based on the number of simultaneous users accessing it. The software licence manager generally tracks and monitors the usage, and any user beyond the maximum concurrency is prohibited.

- **Fixed licence.** This is specific for a single machine, also known as a seat, and is used by a single user at any given time.

- **Named licence.** This is a single licence assigned to an individual user.

- **Virtual users (VUs).** This is generally a defined number of virtual users or 'Vusers' that can use the tool an unlimited number of times within a period. This is commonly used for performance testing tools.

- **Pay-as-you-go licence.** This is a licence based on usage.

Figure 10.1 represents the different licence models widely used in the IT industry.

Software licences are charged and managed these days smartly to provide maximum benefit to the tool users. However, it is important to understand the details of the terms and conditions before procuring the tool. It is strongly recommended to talk to the tool vendors and explore how the licence works as well as the infrastructure and professional support required prior to procurement.

Figure 10.1 Software licence models

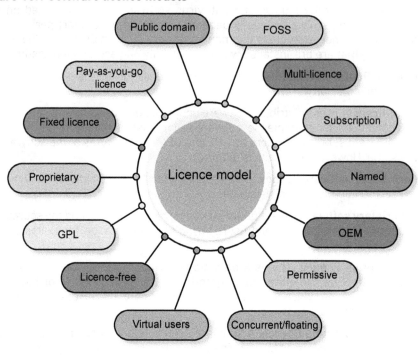

Software licensing is sometimes difficult to fully understand. There are various models used for licence costing and short-term and long-term costs involved. Infrastructure cost is another area to be considered during the costing process. A separate stand-alone licence manager is required to manage the licence in some cases.

Here's an example from the industry:

A leading product development company required a performance and volume testing tool for their DevOps. The tool was expected to be used for daily testing, as well as a final round of full testing. Without a detailed understanding of it, they procured a tool with sufficient virtual users. However, they soon realised that the virtual users were not assigned periodically. The virtual users are charged in a 'pay-as-you-go' manner, that is, once you use a virtual user for a test the overall list is reduced. For example, if you procured 1000 virtual users for a month and used 1000 for your first test, you would end up with no virtual users for any further testing. This was different from the tool licence for many other tools that were widely available. The key message gained from this was that you should thoroughly understand the licence model before procuring a tool. The company decided to procure a different tool based on monthly subscription, with an unlimited number of virtual users for the remaining performance testing cycles.

10.2 LEADING TEST AUTOMATION TOOLS

This section lists testing categories, types and tools widely considered for test automation. Some of these tools are used for more than one type of testing (see Figure 10.2). For example, Selenium can be used for web testing, functional testing, API testing and so on. There are many more tools and testing categories, such as accessibility testing, that are available in addition to those in the list below, which represents the common and leading testing tools:

- API/web services testing
- Unit testing
- Functional and web testing
- Mobile testing
- Test and defect management
- Cross-browser testing
- Performance and load testing
- Security testing
- AI and ML-powered testing tools
- Test framework tools

Figure 10.2 Testing tools

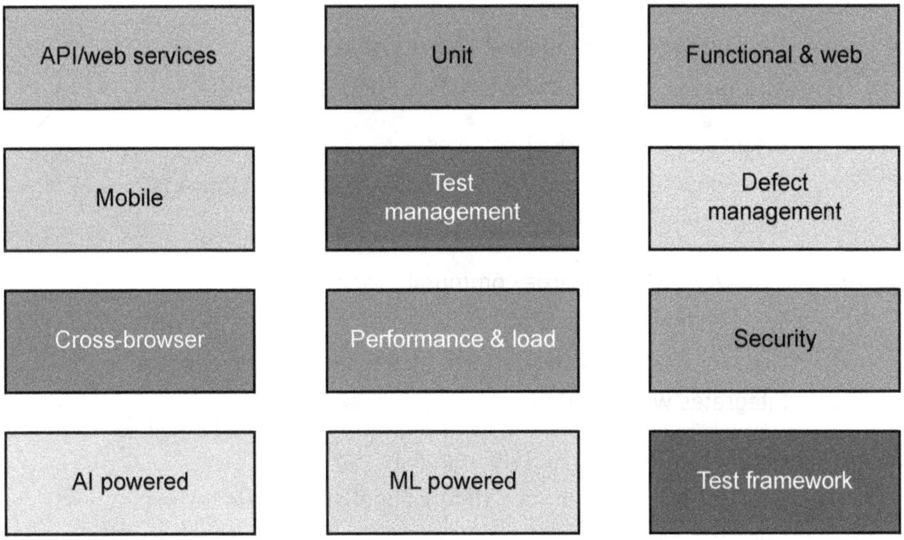

The tables in this section cover a set of leading test automation tools and their key features that are used in the market for testing the categories listed above. There are many more tools available beyond this list, and it is recommended to perform a detailed comparison based on your needs before tool procurement. This section is not a recommendation of any tools over others or an endorsement of tool features.

10.2.1 API/web service testing tools

Table 10.1 covers a set of widely used API/web service test automation tools, and their key features.

Table 10.1 API/web service testing tools

Tools	Key features
SoapUI	Widely used for REST, SOAP, web service and API testing
	Supports record, edit and playback of HTTP communications
	Ease of use including 'drag and drop' and integration with other tools
	Allows the development of your own set of features as SoapUI plugins
	Highly secure and open source
	Used for automated functional, regression, security, API and load testing of web services
	Supports leading technologies and standards
Postman	Popular API testing tool
	Runs on Windows, Mac and Linux
	Facilitates collaboration and sharing of API data
	Allows writing Boolean tests within Postman interface
REST Assured	Framework to test REST services
	Open-source platform with Java domain-specific language
	Good community support
RapidAPI	Widely used with a large community base
	Tests wide variety of APIs
	Used for functional and performance testing of web services
	Integrates with CI/CD tools
Katalon Studio	Supports SOAP and RESTful requests
	Used for API/web services, UI functional and mobile testing
	Supports data-driven approach

10.2.2 Unit testing tools

Table 10.2 covers a set of widely used unit test automation tools, and their key features.

Table 10.2 Unit testing tools

Tools	Key features
JUnit	Open-source unit testing framework
	Supports TDD environment
	Specifically designed for Java programming
	Widely used and popular
NUnit	The unit testing framework is based on .NET
	Supports data-driven tests that can run in parallel
	Popular, and large user community
TestNG	Open-source test automation framework for Java programming
	Resembles JUnit and NUnit
	Supports concurrent and data-driven testing
	Widely used for functional and integration testing
HtmlUnit	An open-source unit testing framework that supports JavaScript
	Framework for testing web applications
	It provides GUI features such as forms, links and tables
	Offers an open-source Java library
	Supports protocols like HTTP, HTTPS

10.2.3 Functional and web testing tools

Table 10.3 covers a set of widely used functional and web test automation tools and their key features.

10.2.4 Mobile testing

Table 10.4 covers a set of widely used mobile and tablet test automation tools, and their key features.

Table 10.3 Functional and web tools

Tools	Key features
Selenium	Open-source functional testing tool
	Popular and widely used
	Used for web application testing
	Accepts test scripts in a variety of languages such as C#, Java, Perl, PHP, Python, Ruby and Groovy
	Good community support
	Compatible with various browsers, Windows, Mac and Linux
	Provides record and playback features
	Integrates with Agile, DevOps and CD workflows
	Integrates with Selenium Grid, which allows cloud-based cross-browser testing using tools such as BrowserStack and SauceLabs
	Supports mobile testing
	A large set of libraries of plugins available
	Provides cross-browser, cross-platform and third-party integrations
	Supports parallel test execution
UFT One/ QTP	Leading functional and web automation tool
	It is licensed software
	Supports Java, .Net, SAP, Oracle, web services
	Supports mobile testing, cross-browser and third-party integrations
	Supports data-driven and keyword-driven tests
	Supports leading test management tools
Tosca	End-to-end functional and web testing tool
	Uses a model-based approach
	Independent of software technologies
	Used for functional and mobile testing
	Provides dashboards, analytics, integrations and distributed executions
	Supports continuous integration and DevOps practices
	Requires minimal technical knowledge
Eggplant	A popular tool for functional, web and mobile testing
	Test cases can run from the command line
	Supports Windows, macOS, iOS and Android

Table 10.4 Mobile testing tools

Tools	Key features
Appium	Open-source test automation tool
	Mobile testing tool for hybrid, native and web-based applications
	It supports cross-platform and works on Mac and Windows
	Creates UI tests and supports code reuse
	Supports test scripts across multiple programming languages, e.g. Java, Python, Ruby, JavaScript, PHP
	Integrates with CI/CD tools
	Integrates with various frameworks and other tools
TestComplete	Provides UI tests for web, desktop and mobile applications
	Allows testing both native and hybrid mobile apps
	It supports testing on multiple mobile platforms
	Supports JavaScript, Python, JScript, C# and C++
	Integrates with various CI/CD tools
	Offers a record and playback feature and supports keyword-driven tests
	Built on top of the Appium open-source framework
Robotium	Open-source library designed specifically for Android UI testing
	Supports native and hybrid applications
	Integrates with CI/CD tools
	Integrates with various frameworks and other tools
Kobiton	Widely used for mobile testing
	Supports Appium
	Integrates with CI/CD tools
	Integrates with various frameworks and other tools

10.2.5 Test management and defect management tools

Test management tools provide features to manage the end-to-end testing life cycle and process. Defect or bug management tools facilitate reporting, tracking, monitoring, closure and metrics of defects and bugs. A suite of test management, defect management, functional test automation and non-functional test automation tools from the same vendor is highly recommended as this helps in data flow across tools, providing a more accurate picture of the quality status, detailed reporting and single centralised licence management. These tools play an important role in test automation as they help to run the tests in a centralised manner, either on a scheduled nightly build or on a continuous integration build.

Table 10.5 provides a limited set of tools for reference. There is a wide variety of open-source and licensed tools available to cater to test management and defect management needs.

Table 10.5 Test and defect management tools

Tools	Key features
ALM	Micro Focus Application Lifecycle Management (ALM) provides features for requirements management, test planning, test case development, test execution and defect management
	Popular with Waterfall projects
	Web-based tool
	Widely used for test and defect management
	Available in on-premises and cloud versions
Azure DevOps	Provides version control, requirements management, automated builds, testing, reporting and release management
	Popular with Agile and DevOps projects
	Widely used for defect management
	Web-based tool and available in on-premises and cloud versions
Jira	Provides features for test, requirement, defect, reporting and release management
	Many of the testing features come from add-on products from third parties with an additional licence fee
	Web-based tool and popular with Agile and Kanban projects

10.2.6 Cross-browser testing tools

Cross-browser testing tools provide features to test web applications on multiple browsers in a short span of time. This is known as browser compatibility testing and validates the solution across Chrome, Firefox, Internet Explorer (IE), Edge, Safari and other browsers.

Table 10.6 covers a set of widely used cross-browser testing tools, and their key features.

10.2.7 Performance and load testing tools

Table 10.7 covers a set of widely used performance, volume and load testing tools and their key features.

Table 10.6 Cross-browser testing tools

Tools	Key features
LamdaTest	Cloud-based cross-browser testing platform
	Supports manual, visual and automated testing
	Validates cross-browser testing on desktop and mobile browsers
BrowserStack	Supports over 2000 different real browsers and devices versions
	Provides features to validate cross-browser and operating system testing of web solutions
	Supports CI/CD
	Supports Windows, Android, iOS and macOS platforms
Browsera	Validates cross-browser layouts and scripting errors for web solutions
	Provides automated browser compatibility testing
	Site crawling feature allows testing of all web pages on a single site

Table 10.7 Performance and load testing tools

Tools	Key features
LoadRunner	Widely used and popular tool
	Licence-based testing tool for load testing
	Large user community available
	Compatible with Windows and Linux
	Works on several enterprise environments
	Supports several types of protocol
JMeter	Open-source tool for load testing
	Widely used with a large user community
	Supports various server types
BlazeMeter	Widely used for load testing of mobile apps, websites and APIs
	Licence-based tool for load testing
Silk Performer	Licence-based testing tool for load testing
	Large user community available

10.2.8 Security and penetration testing tools

Table 10.8 covers a set of widely used security, vulnerability and penetration testing tools and their key features.

Table 10.8 Security and penetration testing tools

Tools	Key features
Netsparker	Security testing tool to check vulnerabilities in web applications
	Popular and widely used penetration testing tool
	Searches for exploitable SQL and cross-site scripting (XSS) vulnerabilities in web applications
	Validates cross-site scripting to SQL injection
Acunetix	Security testing tool to check vulnerabilities in web applications
	Widely used penetration testing tool
	Validates cross-site scripting testing and in-depth SQL injection
	Provides on-premises and cloud testing solution

10.2.9 AI/ML powered testing tools

AI and ML-powered testing tools predict where quality issues are most likely to occur, correlate data, and quickly identify and resolve issues accordingly.

Table 10.9 covers a set of widely used AI/ML testing tools and their key features.

Table 10.9 AL/ML powered testing tools

Tools	Key features
Applitools SDK	Supports all major test automation frameworks and programming languages covering web, mobile and desktop apps
	Integrates with GitHub, Microsoft Azure, DevOps etc.
Eggplant AI	Uses model-based approach to auto-generate test cases
	Increases test coverage by designing test cases-based user behaviour
Retest	Automatically automates tests
	Detects functional differences and visual differences
Mabl	Provides features for test generation and execution
	Provides scriptless tests, cross-browser testing and adaptive auto-healing
	Measures the performance of the application

10.2.10 Test framework tools

There is another set of tools under the wider category 'test framework tools' that are frequently used by the business users and non-technical testers in test automation effort. These enable data-driven testing and keyword-driven testing much more than single test automation tools or libraries. This list includes test frameworks in general that can call other libraries and tools. This category includes tools targeted to support BDD and ATDD approaches.

Table 10.10 covers a set of widely used tools to support business users and non-technical testers, and their key features.

Table 10.10 Test framework tools

Tools	Key features
Cucumber	Widely used for BDD
	Available for most mainstream programming languages
	Uses business facing documentation and is used for ATDD
JBehave	A framework for BDD
	Supports ATDD
Behat	An open-source BDD framework for PHP
	Enables communication between developers, business and testers
	Test scenarios are written using Gherkin
Fitnesse	Supports ATDD
	Widely used to collaborate with business users
	Open-source tool used for web/GUI tests
Serenity	Widely used for ATDD and regression tests
	Integrates with BDD and REST Assured for API testing
	Works as a wrapper on top of Selenium WebDriver and BDD tools

10.3 COST FACTORS

Cost is a key factor in test automation and test automation tool setup. It is important to understand the various costs involved in tool procurement, licence model, annual maintenance, professional services and, most importantly, the infrastructure required to set up test automation tools. Below is a list of cost factors to be considered and explored further during the tool set-up stages:

- Tool infrastructure and setup – Open-source and licensed automation tools setup involves infrastructure costs. Performance testing involves high infrastructure cost and a large volume of virtual users. Client, server, database, reporting and storing need to be considered.

- SaaS, PaaS and hybrid – Cloud and on-premises versions of the tools involve different costs. A cloud-based version involves enabling access to the cloud and configuring the test environment to the SaaS/PaaS platform. This may involve changing the firewall, load balancer configuration, network changes, adding Internet Protocol (IP) addresses to the white list and so on. The security aspects should be considered while accessing the cloud-based testing tools. All the above factors need to be considered for estimating the effort and cost.

- On-premises – On-premises and stand-alone secure environments require additional cost to download the tool to a separate media and physically install the tool to the environment. The test team may require special privileges and access to the environment to deploy the tool. This involves building and accessing the physical or virtual infrastructure. The time to coordinate with the vendor, infrastructure team, scheduling the downtime to do the upgrade, rolling back if something goes wrong and so on all need to be considered.

- Detailed design – An automation tool is another software application in the environment, and any tool in the environment may require a detailed design and approval from the engineering team before installation and deployment. Sometimes it requires a change board to approve the installation. This will involve additional effort from the test team and other teams such as technical architects or change management.

- Procurement – In a consulting or SI environment, it is recommended to procure the automation tool through the SI as they have good expertise in procuring tools for many customers and long-term partnerships with the tool partners. This helps to procure the tool at a considerably reduced rate. However, there is a service fee charged by the SI.

- Training – There is a learning cost involved for many tools, and tool vendors charge for training and installation support.

- Professional services – Ongoing support and maintenance from the tool vendor.

- Licence model – Understand how the licence costs work, for example a leading performance testing tool in the market has a 'use and lose' model as every load testing reduces the Vusers from the total numbers procured.

- Renewal – The renewal charge is another factor to be considered for tool selection and cost.

It is important to consider all the above cost factors before procuring the test automation tool. You may need to contact an existing tool user and the respective vendor to understand the costs involved in test automation tools.

10.4 SUMMARY

This chapter summarised licence models for tools, the most common and widely used tools available in the market at the time of writing and the cost involved in test automation tool setup. This is a key area in ROI calculation and tool selection. The next chapter further explores coding and programming languages used for test automation.

11 TEST AUTOMATION AND PROGRAMMING LANGUAGES

In the last chapter, we covered leading automation tools, their features and various costs involved in procuring and setting up these tools. Customisation of these tools makes them more suitable for organisational needs, and programming is required for customisation. In this chapter, we will explain programming and scripting in test automation at a high level and introduce you to some leading programming languages. These languages will help you in utilising most of the tools described in the previous chapters and creating scripts for specific needs. This chapter is not intended to explain how to do programming. Its purpose is an introduction to automated testing and programming languages. This chapter also addresses coding techniques and candidates for test automation.

Coding, scripting and programming are used to represent the same activity in automated testing.

Coding means using a computer language to create a set of instructions for the computer to behave as desired.

Scripting creates a set of commands that automates the execution of tasks. Scripting languages are often interpreted rather than compiled. Scripting languages are used to create scripting instructions.

Programming comprises a set of commands to produce a list of outputs. Programming languages are used to create programming instructions and implement algorithms.

A set of codes form a script, and a set of scripts form a program.

Programs need compiling before running, whereas scripts need interpretation. Compiled programs generally run swifter than interpreted programs as they compile a code in a complete set, and an interpreter interprets it line by line.

Scripting languages are usually coded in one language and interpreted within another program; for instance, JavaScript is included within HTML webpages and is interpreted by the web browser. Scripting languages are less code-intensive as compared to

146

programming languages. In general, there is a difference between programming and scripting, but they are similar in many ways and hard to differentiate.

Programming languages are grouped into five different generations (Figure 11.1):

- First generation languages (1GL), for example machine-level programming languages
- Second generation languages (2GL), for example assembly languages
- Third generation languages (3GL), for example C, C++, Java and JavaScript
- Fourth generation languages (4GL), for example Perl, PHP, Python, Ruby and SQL
- Fifth generation languages (5GL), for example Mercury, OPS5 and Prolog

Figure 11.1 Generations of programming languages

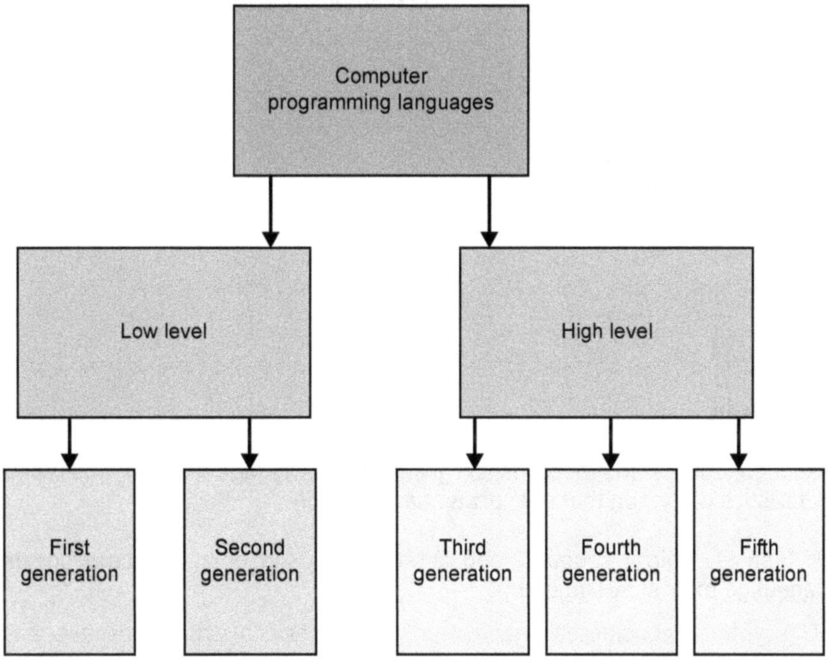

Scripting languages are grouped into two categories:

- Server-side scripting languages
- Client-side scripting languages

Most of the widely used automation tools use programming along with scripting for automated testing. Scripting is extensively used in non-functional automated testing, such as build validation. Perl, Python, Ruby, C# and Java are examples of programming languages, and Jscript is an example of a scripting language.

Good knowledge of programming and scripting is essential to be successful in automated testing.

11.1 PROGRAMMING LANGUAGES FOR TEST AUTOMATION

This section describes the key features of some leading programming and scripting languages used for test automation, with the aim of introducing you to different programming languages that can support the test automation need for your product or organisation.

Automated testing requires proficiency in programming languages and automation frameworks. The first step is to ensure that the SUT, test automation tool for automation and the programming skills in the test team are aligned. The commonly used programming languages to assist in test automation and their key characteristics, which are covered in this section, are:

- Python
- Perl
- Java
- C# (.Net)
- Ruby
- PHP
- JavaScript

11.1.1 Python

This section covers the popular programming language Python and some key characteristics of Python that support test automation:

- Python is a highly popular and extensively used open-source programming language for test automation.
- The syntax is considered easy to learn, and there is a strong presence of Python community at various technical levels, including end-users and programmers.
- Taught in the school curriculum in many countries, which creates a large pool of technical resources with good knowledge of Python coding.
- Python libraries reduce the amount of code writing necessary.
- High portability for the SUT and test automation suites developed in Python and it is compatible with many operating systems.
- Selenium and Appium libraries are available for Python.

11.1.2 Perl

This section covers Perl and some key characteristics of Perl that support test automation:

- Perl is an open-source and commonly used language for test automation.
- It is ideal for web programming, encryption and there are web-specific modules available in Perl.
- Highly recommended for GUI development and system administration.
- Supports object-oriented, procedural and functional programming.
- Perl supports text manipulation, Unicode and is quickly extendible.
- Used for embedded programming.
- Database integration and C/C++ library interface available in Perl.

11.1.3 Java

This section covers the object-oriented programming language Java and some key characteristics of Java that support test automation:

- Java is owned by the Oracle Corporation.
- It is extensively used and accepted within the development community.
- JUnit is a popular unit testing framework based on Java and is compatible with the Selenium testing tool.
- Many open-source test automation frameworks are developed using Java.
- Java follows the 'Write Once, Run Anywhere' principle, and it provides cross-platform compatibility.

Write Once, Run Anywhere (WORA) or Write Once, Run Everywhere (WORE) was a slogan created by Sun Microsystems to highlight the cross-platform benefits of the object-oriented Java programming language. It is the ability of the programming language to run on most common operating systems (OSs) without any changes.

11.1.4 C#

This section covers the C# language and some key characteristics of C# that support test automation:

- C# is based on object-oriented programming, created and supported by Microsoft Corporation.
- Extensively used for test automation and is a popular language using the .NET framework.

- C# is well suited for SUT based on Android, iOS, Windows and macOS platforms.
- It is compatible with Selenium testing tools and is effective for cross-browser testing.
- There are many test automation frameworks available in C#.

11.1.5 Ruby

This section covers Ruby and the key features of Ruby that support test automation:

- Ruby is an open-source language and is used for test automation of web applications.
- Ruby is popular among the student community due to its friendly syntax and flexible object-oriented architecture.
- It is popular for automated browser testing and considered easy for learning and implementing.
- It is compatible with the Selenium testing tool and is effective for cross-browser testing.

11.1.6 PHP

This section covers the PHP language and some key characteristics of PHP that support test automation:

- PHP is a server-side scripting language and is used for web development.
- It is used to manage dynamic content, databases, session tracking and so on.
- It is widely used in test automation.
- It is integrated with popular databases, such as Oracle and Microsoft SQL Server.
- There is active user community support available for PHP.
- PHP is embedded in HTML.

11.1.7 JavaScript

This section covers the scripting language JavaScript (JS) and some key characteristics of JavaScript that support test automation:

- JavaScript is a popular accepted scripting and markup language. It is used for front-end development and testing.
- It is commonly used for automated testing, and there are a wide range of test automation frameworks available in JavaScript.
- JavaScript, along with Selenium, is extensively used for automated browser testing.

11.2 CODING OR SCRIPTING METHODS AND TECHNIQUES

In the previous section, we addressed different programming languages commonly used for test automation and coding. Test automation inevitably requires the use of scripting to enhance the test scripts. Coding and scripting can be used in several different ways, and this section describes coding and scripting at a high level. This section should be read in conjunction with Chapter 5 as the widely used test automation frameworks are developed by using some of the methods below. Each of the methods in this section has its own advantages and disadvantages.

11.2.1 Linear scripting

In linear scripting, the test actions mimic the actual user actions that would be performed while using the application. These scripts are usually generated by a tool that provides a 'capture and replay' facility. This performs by capturing the user actions and system responses and recording them in an appropriate scripted format. The scripts can then be re-run for testing the SUT.

Advantages:

- Linear scripting requires minimal technical knowledge – The initial learning effort and development effort is low and can be introduced very quickly.

- No programming skills are required since it is click and record – This helps the non-technical business users perform linear scripting.

- Most of the test automation tools provide 'capture and replay' and an inbuilt feature.

Disadvantages:

- The scripts capture all the user actions, including any incorrect ones.

- The script may become inefficient due to repetitive user actions, for example repeated browser back button usage.

- The tests are dependent on the user interface as any minor changes in the user interface will halt further tests.

- The scripts are only applicable to the recorded set of actions and cannot be used elsewhere.

- The linear scripts lack error and exception handling, and any failure in test execution will stop the remaining tests.

- Any change to the SUT may impact all the test scripts, and all the recorded scripts may need to be updated.

- Some record and replay tools use proprietary scripting language; this can make the script difficult to understand, maintain and modify.

11.2.2 Structured scripting

Structured scripting is widely used for coding and test automation. In structured scripting, existing tests are used as the basis for creating two-tier test automation scripts. The first tier is a high-level script outlining the basic steps. The second-tier scripts are more detailed procedural steps that perform the required subroutines. The test tool is programmed with a list of high-level tests to be run. During execution, these will automatically call the required low-level scripts. This is illustrated in Figure 11.2.

Figure 11.2 Structured scripting

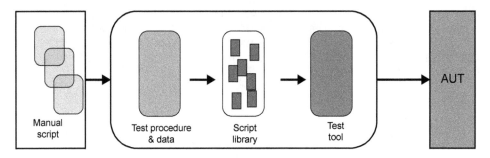

| Manual script | Test procedure & data | Script library | Test tool | AUT |

Advantages:

- The control scripts are structured in two levels and are easy to understand, execute and maintain.
- It is easy to define and develop the second tier and other variants of tests once the initial scripts have been created.
- The shared scripts can reduce maintenance costs.

Disadvantages:

- More effort is required, at the beginning, to design and set up the main and shared scripts.
- Greater technical and programming skills are required compared to linear scripting.
- The script library must be well maintained for future enhancement.

11.2.3 Data-driven scripting

Data-driven scripting is an extension of the structured scripting explained in the previous section, and is where data values are embedded into the high-level scripts. Data files can be stored in a spreadsheet or a database.

Data-driven scripting can lead to many scripts to achieve the required set of tests for range testing. It is a further extension to structured scripting to extract the required data into separate data files. The high-level control script is modified in order to read data items from the data file. Therefore, one script can implement several tests by reading data from different data files. This reduces the script maintenance required per test.

Advantages:

- Reduced build costs since it is faster and easier to build similar test procedures.
- Easy to create many variations of the basic test but using different data.
- The external data file can be created and maintained by the business as this provides an additional guarantee to the test automation coverage.

Disadvantages:

- More initial effort is required to ensure the various files will all operate together correctly (and to meet future needs).
- More initial effort is required to program the main and low-level test scripts.
- Requires a higher level of programming skills.

11.3 CANDIDATES FOR TEST AUTOMATION SCRIPTING

It is rarely feasible to automate all test cases and tests, as some tests will be difficult to automate or will return low benefits when compared to the effort required. This section lists a set of questions to be asked before selecting the tests, feasibility of features and identifying requirements for test automation.

- Is the test case repeatable? It is good for automation if it is used very often. Tests used once or infrequently are not suitable for automated scripting as the ROI will be low.
- Does the test require any manual intervention? If the tests require manual intervention to complete they are not a good candidate for automation, for example many scientific applications.
- Could all test preconditions be set up automatically? If so, this is a good candidate as it requires less manual intervention, for example setting up test data automatically for testing.
- Are the expected results dynamically changing? These tests are difficult to automate and require a lot of scripting and coding effort, for example Global Positioning Systems (GPSs).
- Does the SUT change dynamically, for example using data feeds? They are difficult to automate because the expected result changes constantly, and successful implementation requires a lot of effort, for example accessing the data feed at the source to match with the result on the front end.
- How easy is it to set up test data for the test case? In some cases, test data need to be set up manually. This will increase the effort for test preparation, reduce the reuse of automated testing and the ROI will be low for test automation.

- Are they UI/UX tests? User experience, also known as 'look and feel', tests require manual intervention to validate the results and are difficult to script. They are not considered good candidates for automation.

- Could a clear pass or fail criterion be defined? It is important for the tool to validate the actual result against the expected result. It is difficult to script dynamically changing or ambiguous outcomes for the same inputs. Many test automation tools provide tolerance in the expected result. However, this can impact the accuracy of the testing.

- Could the test case be manually validated with less effort? These tests may not be good candidates for test automation as scripting and test automation delivers a lower ROI compared to manual testing.

- How quick would it be to update the test if the system functionality changes? This is a key factor for scripting as it takes a lot of effort to maintain the scripts.

- Is it difficult to code or script? Avoid these tests as automated scripts can be 'flaky' and difficult to debug.

A test script is a set of instructions for automated testing using a scripting or a programming language.

11.4 DEVELOPING YOUR FIRST SCRIPT

A systematic and disciplined approach is always useful for test automation and automated testing. It is difficult to introduce all the automation in a single exercise, and easier to introduce automation in a number of disciplined steps and stages. This section provides a set of examples to commence test automation, automate the first script and extend the first script to a test automation suite.

- **Example one.** Start with a simple record–replay tool to create an initial set of automated scripts and then use scripting or coding techniques to enhance the 'record and replay' script to develop a more powerful script; for example enhance the 'record and replay' to a data-driven script and develop a test suite that can be used for validating a large volume of inputs.

- **Example two.** Using an existing automation framework or suite helps in building the automation experience slowly and avoids delivery pressure. Extending the framework libraries or adding new automated tests to an existing framework supports the learning process.

- **Example three.** Start with a SUT that is in a more stable phase, for example a maintenance phase, where a large number of regression tests might be the first candidates for scripting and test automation.

- **Example four.** Start with an existing product for developing the first script for test automation as the SUT changes are minimal.

Below there are a few alternative options to commencing test automation if support from the leadership and organisation are minimal because of other priorities and commitments.

- Open-source tools
- Record and playback
- Existing test suite

11.4.1 Open source

The automated script needs to be reliable, robust, maintainable, easy to debug and scalable to multiple environments. Open-source tools are a good option as they are cost-effective, require minimum hardware or infrastructure and have massive support available from the community. The first step to using open-source tools is to download the tool, install it and select a simple functionality to automate. Most of the open-source tools and languages provide a beginners' guide to starting automation.

11.4.2 Record and playback

Record and playback is another way to introduce test automation in the organisation. It is a type of automated testing where the tool records the activity of the user and then imitates it while playing back. Most of the widely used tools and frameworks support this feature. Record and playback provide a basic script, which can be enhanced, for example a simple record and playback script for login functionality can be enhanced with a data-driven approach to extend 100 per cent test coverage.

11.4.3 Existing test suite

Another good approach is to start with the existing regression pack. Many organisations and projects continuously use existing regression tests with minimal changes. Over a period, an automated regression pack becomes toothless and obsolete due to lack of maintenance, although part of these scripts can be useful after modification. The obsolete regression pack is a good place to reintroduce automated testing cost-effectively.

11.5 SUMMARY

In this chapter we have looked at the role of programming or scripting languages and coding in test automation. Scriptless tools are now widely used; however, coding cannot be ignored as it enhances the ability of test automation. This chapter also discussed different coding methods and how to start automated testing if you are new in this area. This chapter helps test managers and Scrum masters to introduce test automation in their organisation.

The next chapter will cover test automation framework design and how to start with automating scripts. However, not all organisations and projects approach automation systematically and many organisations lack budgetary support for test automation. Nevertheless, test automation leaders can still introduce automation with minimal resource and budget support.

12   AUTOMATION FRAMEWORK DESIGN AND DEVELOPMENT

This chapter is essential for test automation engineers, architects, project managers, Scrum masters and everyone involved in test automation framework design and development. This chapter explains how to approach a test automation framework by using a spreadsheet for data management. This framework is based on the philosophy of using libraries for validation and reporting.

A framework is defined as a set of rules and guidelines, such as best practices and coding standards, that can be followed in a systematic way that ensures the delivery of targeted results such as increased code reusability, higher portability and reduced script maintenance costs. We discussed different test automation frameworks in Chapter 5 and how they work. The right framework needs to be selected based on an organisation's current and future needs. The essential criteria for any automation framework are its scalability, maintainability, and code that can be quickly understood, easy to debug and rapidly developed.

The primary objective of the framework is to maximise reusability of common components and libraries while minimising maintenance of the framework and automated tests. The framework is expected to provide the following benefits:

- Reusability of functions and libraries

- Quick test execution and low learning curve

- Low maintenance of the overall framework and automated test scripts

- Repeatability of regression tests on different environments

- Quick enhancement of the framework with new features

One key aspect of the test automation framework is that it is designed and developed with the intention of future use. The framework should be 'generic', meaning that it should be compatible with more than one application and product.

The first part of this chapter explains some common components and terminologies used in any standard test automation framework. The latter part of the chapter explains how to design and build a test automation framework.

A test automation framework is a programming framework that includes a comprehensive set of guidelines, standards, software packages, and testing tools to provide a foundation structure for developing a test automation suite.

12.1 TEST AUTOMATION COMPONENTS' DEFINITIONS

In this section, we explore the definitions of some common phrases, terms and activities in test automation and relating to test automation frameworks. The purpose of this section is to familiarise you with the common terminologies in test automation (Table 12.1).

Table 12.1 Common definitions

Term	Definition
Editor	An automated test script is a set of instructions created by the tool or manually created by a human to validate the SUT. Automated scripts can be written in either a programming language using an editor or created by a testing tool. Editing is modifying the scripts using an editor.
Object repository	An object repository is a collection of objects and properties taken from the SUT for a testing tool to recognise the corresponding elements.
Library	A programming library is a suite of helper functions that another program can make use of. This may include common functionality, subroutines, classes, values or type specifications.
Classes	In object-oriented programming, a class is an extensible program for creating objects.
Exception handling	This is the process of a program responding to exceptions or unexpected behaviour.
Recovery scenario	This contains generic functions to handle exceptions at runtime.
GUI objects	An object has a state (data) and behaviour (code) and is an abstract data type with the addition of polymorphism and inheritance. GUI objects are the same, representing GUI elements such as text boxes, buttons and check boxes.
Configuration management	Configuration management is the tracking and controlling of changes in software including version control. Configuration management is essential to maintain software, system and document integrity and consistency once the solution has been developed and released by multiple teams or in multiple environments across a wide range of end-user communities.

Table 12.1 (Continued)

Term	Definition
Deployment	Software deployment is the process of making the application available on a target place: test server, production environment, mobile device etc.
Test automation script	A test automation script is a set of instructions that will be performed on the SUT to validate that it works as expected. Good test scripts are small, maintainable and easy to understand with swift execution of steps. They are independent of each other and run with good exception handling.
Code review	Code review or static testing is a systematic check of another's code for mistakes based on standards and guidelines.
Debugging	The process of finding the cause of a problem in a software code.
Utilities	Contain the functions required at a generic level; these are not application specific.
Verifications	Contain the validation functions that are application specific.

12.2 BUILDING A TEST AUTOMATION FRAMEWORK

This section explains how to build a test automation framework. The process and components of a test automation framework can be better understood by going through a real case study of one developed for a public-facing web-based healthcare portal.

This test automation framework was created to last a decade to cater to functional and regression needs. The test automation architect is the key person responsible for developing and maintaining the test automation framework. They define the coding standards and train the test automation engineers to write automated test scripts.

The key steps to developing this framework were:

1. Framework requirements and scope
2. Test automation framework approach
3. Tool selection
4. Phases of the automation
    a. Initial phase
    b. Development phase
    c. Final phase
5. Defining framework folder structure
6. Complying with guidelines and standards
7. Test script and execution

The above steps will be explained below in the context of the healthcare portal.

12.2.1 Framework requirements and scope

The SUT was a web portal for a multimillion user base that helps users to make choices about health, from lifestyle decisions to the practical aspects of finding information when needed.

This section provides an overview of the scope of test automation – what is automated and what is not. The focus was mainly on the overall need for functional and regression testing. The requirement was the test automation framework development, creation of the regression suite and execution of the created automated scripts.

The framework was expected to support the multiple browsers, operating systems, mobile platforms and environments as shown in Table 12.2.

Table 12.2 Framework platform requirements

Environment	Operating systems	Browsers	Mobile platforms
Live	Windows	Chrome	iOS
Pre-production	macOS	Edge	Android
End-to-end		Safari	
Integration testing		Firefox	
System testing		IE	
Development			

The framework and the test suite were expected to be independent of any environment and build. The following functionalities and features were in scope:

- Content management
- Large set of test data validation
- Backend systems

The following functionalities and features were not in the scope of this framework due to the high level of effort required for minimal returns.

- Videos, audios and embedded flash testing
- User interface and user experience

12.2.2 Test automation framework approach

A test automation framework approach or plan is a document describing the approach to the automated testing and framework. This is the top-level plan generated and used by the test automation team to direct the test effort. This document describes 'how' automated testing is implemented for the solution and 'when'. This provides a summary of the test automation activity and responsibilities specific to the life cycle of the framework. The key steps to forming a test automation framework approach or plan are:

1. Identifying the functional areas or products for automation, for example web services, healthcare.

2. Analysing the functional requirements and test cases, for example review the SUT features and test scenarios for the suitability of test automation.

3. Verifying the feasibility of automation such as the compatibility of the SUT with the selected tool, for example proof of concept to ensure that the proposed tool is compatible with the SUT and is fit for purpose.

4. Identifying the manual tests that can or cannot be automated, for example tests involving frequent manual intervention or UI/UX.

5. Classifying the test cases or requirements into various levels or priorities based on business needs, for example critical features, frequently executed tests.

6. Developing the test automation framework, for example framework libraries.

7. Automating the level one or priority one test cases and requirements for the initial set of tests. The time-consuming manual tests can be expedited and added to this list as well, for example regression tests or build validation tests.

12.2.3 Tool selection

This section addresses the test automation tool selection criteria for the healthcare portal. The tool selection process is covered in Part One of this book. The test automation tool for the healthcare portal was selected based on the following factors:

- The priority was the test automation tools that were widely and commonly used for functional test automation. This ensured good support from the vendors, community and the availability of skills in the market.

- Latest version of the tool based on a PoC result. This ensured that the tool was compatible with the SUT and with the latest browsers available in the market.

- The SUT was developed on Microsoft .NET and SharePoint, and the selected tool was flexible and suitable for Microsoft .NET and SharePoint platforms.

- The selected tool supported test script execution on multiple environments and platforms and in batch mode in order to save time and effort.

12.2.4 Phases of the automation

The software test automation framework (STAF) was based on common library functions, which increased the flexibility of the framework. Automation framework development was carried out in three phases: initial, development and final.

Initial phase

The purpose of the initial phase was to plan the test automation framework with activities such as understanding the solution, analysing the SUT, setting the criteria for candidate selection for tests, defining scripting standards, following guidelines and formulating a test automation framework plan.

The following activities were part of the initial phase:

1. Understanding the architecture and functionality of the SUT.

2. Identifying the application functional areas and tests to be automated.

3. Developing the scripting approach.

4. Agreeing on the automation framework approach and plan.

5. Designing test automation framework, scripting standards and guidelines.

Development phase

The objective of this phase was to create the automation framework based on the plan developed in the initial stage. At the end of this stage a test automation framework and initial set of test scripts were ready for the product.

The following activities were part of the development phase:

1. Defining, creating and collecting test data.

2. Building common libraries of the framework.

3. Creating test automation scripts for prioritised tests.

4. Executing the trial or initial test scripts (dry run).

5. Analysing the test results from the test execution and test reports created by the framework.

6. Peer reviewing the test scripts.

7. Preparing the user manual for framework, script maintenance and execution.

8. Sharing the framework and test scripts with the wider testing, development and support teams.

9. Enhancing the test scripts for additional features and products.

Once the test execution is completed, the next step is to review the test results and identify functionality or components that experience a relatively high number of failures. The result analysis summarises whether it requires additional test effort to update or enhance the tests scripts and framework. Test results and reports generated from the analysis can confirm the executed test scripts are suitable to identify defects in the SUT. It is the last phase of test automation framework development and thereafter the test reports are shared with all stakeholders.

Figure 12.1 Test automation road map

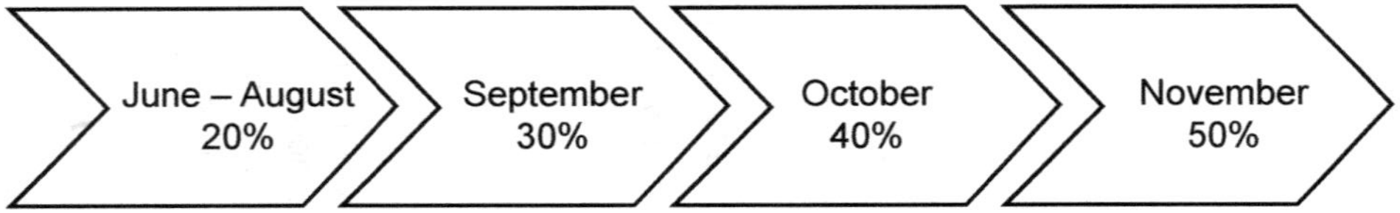

June – August 20%	September 30%	October 40%	November 50%

Regression Test Automation Schedule

S.No	Products	Test | Automation Coverage in %				Level1	Level2	Percentage can be automated	Percentage cannot be automated	Features cannot be automated
		June - August (25%)	September (10%)	October (10%)	November (10%)					
1	Product 1	35% (Level 1)	25% (Level 2)	25% (Level 2)	15% (Level 2)	Details	Details	75%	25%	Details
2	Product 2	30% (Level 1)	15% (Level 2)	20% (Level 2)	15% (Level 2)	Details	Details	80%	20%	Details
3	Product 3	30% (Level 1)	25% (Level 2)	20% (Level 2)		Details	Details	75%	25%	Details
4	Product 4	35% (Level 1)	10% (Level 2)	10% (Level 2)	10% (Level 2)	Details	Details	65%	35%	Details
5	Product 5	70% (Level 1)	10% (Level 2)	To be confirmed		Details	Details	80%	20%	Details
6	Product 6	40% (Level 1)	15% (Level 2)		15% (Level 2)	Details	Details	70%	30%	Details
7	Product 7	30% (Level 1)	30% (Level 2)	10% (Level 2)		Details	Details	70%	30%	Details
8	Product 8	30% (Level 1)	20% (Level 2)	20% (Level 2)		Details	Details	70%	30%	Details
9	Product 9	To be confirmed	20% (Level 1)	25% (Level 2)	15% (Level 2)	Details	Details	60%	40%	Details
10	Product 10	To be confirmed	20% (Level 2)	20% (Level 2)	20% (Level 2)	Details	Details	60%	40%	Details
11	Product 11	To be confirmed	To be confirmed		40% (Level 1)	Details	Details	40%	60%	Details
12	Product 12	To be confirmed	To be confirmed	25% (Level 1)	25% (Level 2)	Details	Details	50%	50%	Details
13	Product 13	To be confirmed	20% (Level 1)	15% (Level 2)	15% (Level 2)	Details	Details	50%	50%	Details

Final phase

The final stage was to transition the automation framework, scripts and related documents including how to set up the framework, user manual and so on to the ongoing support team.

The following activities were part of the final phase of the healthcare test automation framework:

1. Baselining the framework, automated scripts and documentation.

2. Defining ongoing maintenance.

3. Defining a road map.

Figure 12.1 presents the road map for the framework.

12.2.5 Defining framework folder structure

The folder structure shown in Figure 12.2 represents test data, common libraries, exceptions and automated scripts of the framework. This structure can be further expanded with more reusable components and libraries subject to the automation needs, for example 'Reporting'.

Figure 12.2 Framework folder structure

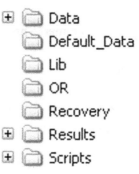

The folders are explained below:

- Data folder contains the test data, which are maintained in a spreadsheet.
- Default_Data contains files that are saved from the SUT.
- Lib folder contains the reusable functions.
- Object repository (OR) contains data related to the objects' descriptions of the SUT stored.
- Recovery folder contains generic functions to handle exceptions at runtime; all the recovery scenarios are stored.

- Results folder contains test execution result files for every execution of the scripts stored.

- Scripts folder contains test scripts.

The framework could have many more additional components such as:

- Global variables – Environment file can be located here.

- Reporting – Libraries for reporting purposes can be available here.

- Utilities – Functions required at a generic level, and not application specific will be available here.

- Verifications – Contains the validation functions that are application specific.

For the framework created for the healthcare application, the common functions and variables were maintained in the Lib folder. Test data were maintained using Microsoft Excel files and were stored in the 'Data' folder. The test data could have been directly used from the database.

Figure 12.3 provides another example of a folder structure for an ecommerce project that used BDD SpecFlow (a .NET open-source framework for behaviour-driven development, SpecFlow helps BDD to turn specifications into executable code).

Figure 12.3 BDD SpecFlow for an ecommerce project (example)

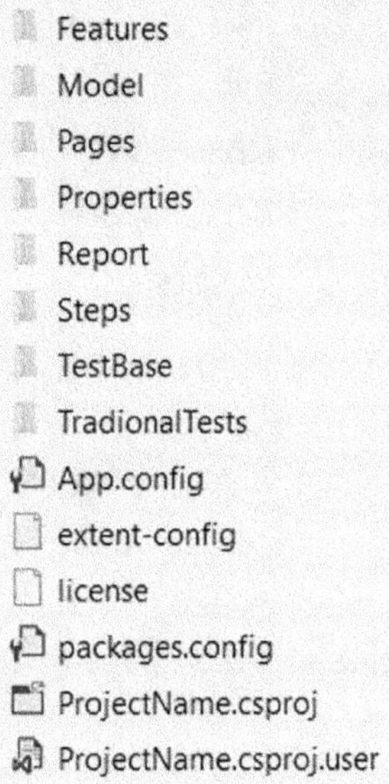

       Features

       Model

       Pages

       Properties

       Report

       Steps

       TestBase

       TradionalTests

       App.config

       extent-config

       license

       packages.config

       ProjectName.csproj

       ProjectName.csproj.user

The folders are explained below:

- Features – Documents that contain Gherkin scripts and other relevant details. Feature files allow the building of test packs for features and activities. This helps in creating a set of regression tests that can be maintained more easily than individual stories.

- Model – The model classes represent domain-specific data and business logic.

- Pages – Page Object model is an object design pattern, where web pages are represented as classes, and the various elements on the page are defined as variables of the class.

- Properties – Programming language properties such as .Net or Java.

- Report – Execution reports are stored in this folder.

- Steps – Step definition is a piece of code with a pattern attached to it or, in other words, a step definition is a Java method in a class with an annotation.

- TestBase – This folder contains the files, browser selection classes, web extensions and so on.

Refer to Appendix B for the folder structure for a framework created in Selenium and C#.

12.2.6 Complying with guidelines and standards

The automation scripts for the healthcare test automation framework were developed based on coding standards, guidelines and naming conventions (see 'Further Reading' for more detail). A proper naming convention based on the agreed standards and guidelines was followed for variables, functions and objects. Comments were specified in the code wherever necessary for ease of later maintenance. A proper indentation is used for a clear understanding of the script. The indentation refers to the space at the beginning of a code line.

12.2.7 Test script and execution

The test automation framework for the healthcare portal was developed by following a hybrid automation approach. This was a combination of a data-driven and a keyword-driven approach. All the test scenarios were built on the data-driven approach to broaden the scope of testing by using various sets of test data (parameterisation).

The execution was initiated from the 'application sheet'. This sheet maintains data related to environment, browser and application functionality. Figure 12.4 provides an example of the application data sheet from the healthcare test automation framework.

Figure 12.4 Application data sheet

Application	Environment	Run	Application Path	Browser
Application 1	Production/Live	Yes	Https://www.Application1xxxxxx1.com	Chrome
Application 2	Production/Live	No	Https://www.Application2xxxxxx2.com	Edge
Application 3	Production/Live	No	Https://www.Application1xxxxxx1.com	FireFox
Application 4	Pre-Production	No	Https://www.Application2xxxxxx2.com	Safari
Application 5	Pre-Production	No	Https://www.Application1xxxxxx1.com	Chrome
Application 6	UAT	No	Https://www.Application2xxxxxx2.com	Edge
Application 7	UAT	No	Https://www.Application1xxxxxx1.com	FireFox
Application 8	SIT	No	Https://www.Application2xxxxxx2.com	Safari
Application 9	Development	No	Https://www.Application1xxxxxx1.com	Chrome

The application areas or modules were maintained in a separate sheet based on the category. This sheet was used to define the scope of any regression testing. The regression test suite was designed by updating the sheets with relevant values. The application areas or modules could be selected through the data sheet with yes or no values.

Figure 12.5 provides an example of the application areas/modules data sheet from the healthcare test automation framework.

Figure 12.5 Application areas/modules data sheet (regression suite)

Product	Run
Product 1	Yes
Product 2	No
Product 3	No
Product 4	No
Product 5	No
Product 6	No
Product 7	No
Product 8	No
Product 9	No

It is necessary to update a data sheet and execute the test if a new environment is added the requirement.

Figure 12.6 provides an example of the regression test data sheet from the healthcare test automation framework.

Figure 12.6 Data sheet (regression suite)

Title	Link 1	Link 2	RSSLlnk	RUN
Syndication title 1	Link	Link	Link	Yes
Syndication title 2	Link	Link	Link	No
Syndication title 3	Link	Link	Link	No
Syndication title 4	Link	Link	Link	No
Syndication title 5	Link	Link	Link	No
Syndication title 6	Link	Link	Link	No
Syndication title 7	Link	Link	Link	No
Syndication title 8	Link	Link	Link	No
Syndication title 9	Link	Link	Link	No

After every test execution, a report was generated based on the selected environments, browsers, operating systems and application functionalities. This report was generated in HTML format by the reporting scripts created as part of the framework, stored in the results folder and was scheduled to be sent automatically to various stakeholders by email.

Figure 12.7 provides an example of the test report from the healthcare test automation framework.

Figure 12.8 provides an example of the test execution summary from the healthcare test automation framework.

Figure 12.9 provides an example of the detailed result from the healthcare test automation framework.

This section has provided an overview of the structure and components of a real-life hybrid test automation framework for the healthcare portal. It is to be noted that the creation of a framework was subject to various factors such as tool availability, SUT, skills, ROI and scope of automation.

12.3 SUT ARCHITECTURE

Good knowledge of the SUT and its architecture is important for test automation framework design. The solution architecture design document or full system architecture document is a good place to go to understand the SUT architecture. The physical and logical architecture design documents provide insight into server types, network architecture, storage requirements, interfaces, network design, system components, related components, sub-components, data architecture, security architectures, the

Figure 12.7 Test report

Sub Test Type	Status	Reasons for failures
Smoking	Passed	
Tools	Passed	
Tools---Audio	Passed	
Video	Passed	
Live Well	Passed	
Pregnancy	Passed	
Sexual health	Passed	
Alcohol	Passed	
Allergies	Passed	
Drugs	Passed	
Fitness	Passed	
Good food	Passed	
Lose weight	Passed	
Mental health	Passed	

Figure 12.8 Test execution summary

Workitem number	Title	Status	Failed steps	Total steps
693	Document Library Public Views	PASSED	0	7
732	Task List	PASSED	0	16
730	Calender List and WebPart	FAILED	2	5
728	Announcement List and Webpart	PASSED	0	20
694	Checked out to View	PASSED	0	4
707	Picture Library	PASSED	0	11
680	Quick Launch Bar	FAILED	4	11
764	Content Type Hub	PASSED	0	15
690	Document Library	PASSED	0	45

Figure 12.9 Detailed result

WorkItem number	Title	Execution Start Time
693	Document Library Public Views	10:01:42

Execution time	Description	Status	Steps
10:01:48]	Check, Homepage exists	Pass	1
10:01:55]	Check, Cliked on the Workplace Documents in the Document library item	Pass	2
10:01:57]	Check, Workplace Documents' list available	Pass	3
10:02:00]	Check, My Latest Documents view exists on the Public view	Pass	4
10:02:06]	Check, Documents by Author view exists on the Public view	Pass	5
10:02:12]	Check, Documents by Content Type view exists on the Public view	Pass	6
10:02:19]	Check, My Checked out items view not exists on the Public view	Fail	7
	Test End	10:02:19]	

WorkItem number	Title	Execution Start Time
730	Calender List and WebPart	10:02:19

Execution time	Description	Status	Steps
10:02:25]	Check, Homepage exists	Pass	1
10:02:27]	Check Workplace Calender Webpart exists on Workplaces Homepage	Pass	2
10:02:27]	Check Calender Webpart exist with expected name as - Workplace Calendar	Pass	3
10:02:27]	Check Calender Webpart has List Item as - Event1	Fail	4
10:02:27]	Check Calender Webpart has List Item as - Event2	Fail	5
10:02:27]	Check Calender Webpart has List Item as - Event3	Fail	6
10:02:27]	Check, Add new event link is available for user to create new event	Pass	7
10:02:41]	Check, Quick launch menu exist on the Home page	Pass	8

generic structure of the solution, logical information, capacity and service continuity. This information can be located from the following placeholder documents of the SUT:

- The solution architecture design or full system architecture – Defines the generic structure of the solution, providing the logical and physical information architecture for the selected option, including all related components and interfaces, capacity and service continuity designs.

- Detailed design and physical design – Defines the design of the physical architecture of the selected logical option, server types, network architecture, storage requirements and interfaces. This also includes any network design.

- Security architecture and cybersecurity assessment – Provides insight into the security aspects and risks associated with the implementation of the solution and used subsequently to show that security has been adequately considered. It highlights the threats and vulnerabilities and investigates countermeasures to provide assurance to the business that adequate security has been implemented.

- Component design architecture or low-level design – Provides the detailed architecture for a component. It describes the services a component provides, its sub-components and the supporting architectural views for a component, for example application, data and security architectures.

- IT business and service continuity – Details the instructions required to failover or restore a system or service in the event of failure.

- Test strategy and plan – Defines the strategy for the assurance and testing effort required for the solution. It provides a summary of the test stages and responsibilities specific to the product or project. It may be high level (dependent upon availability of functional and non-functional test requirements) and elaborated in the next phase. This includes all the information necessary to plan and control the test effort for the solution development phase. It describes the approach to the testing of the artefacts and is the top-level plan generated and used by managers to direct the test effort.

It is essential to comprehend SUT architecture for the test automation framework design as any failure to understand application architecture and avoid decoupling components in the automation framework impacts the long-term ROI and the maintenance of the framework. For example, failure to separate the UI of an application from the automated tests will affect all related automated tests due to any minor change to the UI of the SUT.

High-level design and full system architecture are sources of information on how the overall system applications are developed as part of their overarching enterprise architecture, including what technology is used and how these subsystems interact with each other to meet the overall business needs.

12.4 TEST DATA MANAGEMENT

Data management and maintenance is a key part in any organisation. Many automation frameworks facilitate their own data management by gathering data from the business and maintaining them as part of the framework. Test data are the input given to a SUT during test execution and are used to validate both normal and alternative conditions. The positive data (input) are used to verify that functions produce expected results, and the negative data (input) are used to check unusual, exceptional, abnormal or unexpected outcomes.

A process of planning, designing, storing and managing data for a SUT is called software test data management.

One of the key challenges for test automation is managing test data. Test data are key for the application to work well with the automation tool. Lack of accurate and realistic live-like data will be impractical for automation. Test data can be created, or existing test data from different environments can be considered for test automation. However, it is not always possible or feasible to transfer data from one environment to another for technical and data protection reasons.

Test data should be defined and should be available prior to performing automated testing, unlike manual testing data, which can be created during test execution. Defining these data are an integral part of developing both the manual test scripts and automated test scripts.

There are various challenges in defining and creating test data. These challenges should be addressed prior to scripting. The list below describes some key challenges in test data management:

- Data change often – This can cause a particular test to fail and creates a new requirement for additional or new data. The test script can be programmed to accept a level of tolerance to overcome this situation.

- The business does not have the time to verify the data with the automation team – This creates a dependency on the business, which is always a bottleneck when performing automated testing. The automation framework should be programmed to collect the data directly from the source if possible.

- Errors in data imported to the framework – This is a key issue, specifically in transactional test data. The automation framework should be programmed to collect the data directly from the source if possible.

- GDPR and PII data cannot be shared with the test team – Due to the strict GDPR and PII guidelines, it is not often possible to use real-life, personal data for automated testing. The automation framework should be programmed to collect the data directly from the production or a production-like environment if possible. The test automation framework can be installed and configured on the production environment to avoid any data breach. The test execution environment needs to be accredited prior to using PII data.

- Data integrity and data accuracy – This is a key issue when data are created manually or when using a tool for testing. The automation framework should be programmed to collect the data directly from a production-like environment, and the test environment refreshed from the production or production-like environment prior to test execution.

- Environment accreditation – A prerequisite for many secure and financial systems is to use live-like data or data from the production system. The test execution environment needs to be accredited prior to using PII data. The accredited process is generally effort intensive and involves multiple teams.

Test automation starts tackling the data from the planning stage, as waiting until the automated test execution stage is too late and impacts the testing. The solution needs a comprehensive test data strategy in place prior to automated testing. Test automation can also reuse the data from the manual testing or the data created by load and volume testing. Test data can be created in different ways. The list below and Figure 12.10 cover different methods and sources of data for test automation. These are subject to your SUT and business domain.

- Manually created data – For example, use the data from the manual testing.

- Test data creation tools – For example, use test data creation tools such as DATPROF, Informatica Test Data Management or InfoSphere Optim.

- Production data – Use the data from the source or the live system for test automation.

- Test automation tool/framework-created test data – Use the test automation framework to create the test data such as user accounts.

- Purchase – Different types of test data can be purchased such as postal, telephone or email, for example from organisations such as UK Datahouse or Data HQ.

- Existing data – The data used for the previous automated testing cycle can be used again for the next cycle.

Figure 12.10 Test data creation

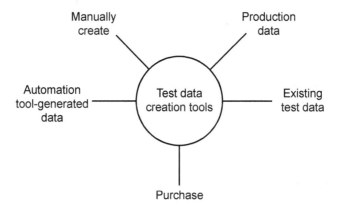

Test data are generally different from production data. However, often production data are used as test data. Test data management tools are widely used to manage data for manual testing and test automation.

Table 12.3 covers a set of widely used test data management tools, and their key features.

Table 12.3 Test data management tools

Tools	Key features
DATPROF test data management tool	A test data management tool for data masking
	Provides synthetic test data generation and a test data provisioning platform
	Manages and refreshes test data across environments from one central location
	Compliance with GDPR
	Integrates and automates various test data processes within the CI/CD
Informatica MDM	Provides automated data subsetting, data masking, data connectivity and test data-generation capabilities
	Provides cloud-based solution
	Manages multiple domains such as customer, product or supplier
CA Service Virtualization (formerly known as LISA)	Simulates the behaviour data
	Creates a virtual data set
	Imports test data from different types of data sources such as Excel sheets, XML and log files
	Automatic data masking protects sensitive data without violating any security policy
IBM InfoSphere Optim	Offers data management from requirements to retirement
	Supports continuous testing and Agile software development
	Provides real-time data testing

12.5 SUMMARY

Test automation makes software testing easier, faster and more reliable if implemented correctly and is essential in today's fast-moving software delivery environment. Usually seen as an alternative to time-consuming and labour-intensive manual testing, test automation uses software tools and frameworks to run a large number of tests repeatedly to make sure an application does not break whenever new changes are introduced. This chapter summarised how to design a test automation framework and components of the framework. This chapter also addressed the importance of SUT architecture and test data management in test automation.

The next chapter addresses another key area in test automation: how to measure the progress, status and success of test automation and KPIs.

13 MEASURING TEST AUTOMATION

Management thinker Peter Drucker has been quoted saying that 'What gets measured gets managed' (Prusak, 2010). This is very relevant for test managers in automation as it is a substantial strategic and tactical investment; hence it needs to be measured correctly. This chapter advises test managers and leaders on how to measure test automation quantitatively on quality terms such as test coverage and metrics.

Metrics are used to measure performance quantitively and provide the key management information for internal and external stakeholders. Software test metrics are used to measure testing activities and track progress and status. KPIs are parameters used to determine the success of business activities. KPIs for test automation are used to measure the efficiency and effectiveness of automated functional testing, automated regression testing and so on. This chapter introduces various metrics and KPIs used for test automation, to be selected based on the organisation's needs. KPIs and metrics need to be quantitative, reflect the business priorities and be agreed up front with the stakeholders with periodic reviews.

Test automation metrics measure the performance, past, present and future, of the automated testing process to determine if it is delivering an acceptable return on its investment. Implementing automated testing is a process, and any metrics chosen to measure improvement (e.g. the test coverage after implementation of test automation) need to take into account the unique aspects of the organisation, market or environment they are being used in. There is no set of 'universal metrics' that will work in every capacity, all the time.

> Test automation metrics are used to measure the performance (past, present, future) of the implemented automated testing.

As metrics are used to measure success, it is important that the KPIs and metrics are clearly defined and agree about what success looks like. The effort and outcome should be clearly measurable for automation metrics. Metrics are also used to determine areas for improvement. Test automation metrics reveal the true state of test automation performance, and they help in making the right decision.

Some key benefits of test automation metrics and KPIs are:

- Test automation metrics indicate effectiveness, including effort savings, comparison to manual testing effort, benefits and the ROI. This helps in making decisions on further investment in test automation.

- Requirement coverage metrics show what features are tested and trace them back to the relevant user story or requirement. Requirement coverage and traceability by test automation provides confidence in how well the solution is tested by test automation and assists decision making.

- Code coverage is a measurement of unit test coverage. Code coverage metrics are widely considered as KPIs for unit testing and automated unit testing. Code coverage metrics provide an indication of the stability of the SUT, effectiveness of automated build and deployment validation.

- Test execution metrics include counting the number of tests passed or failed, giving an overview of testing progress and status. This provides end-to-end traceability, more visibility of software quality, prevents escaping defects, exposes quality risks for Sprint planning and helps to improve quality at the organisational level.

- Various defect metrics indicate the effectiveness of testing, quality of the application and defect trends. Defect trends help to improve quality through understanding the behaviour of defects, defect density, common quality issues, defect arrival rates, defect fix trends and so on. Defect metrics also help in achieving cost savings by preventing defects.

The measurement parameters below are used to define various automation test metrics suitable for the organisation. Comparing some of these parameters against manual testing will show how test automation is performing against manual testing. It is essential to understand the organisation's needs and use combinations of measurement parameters to define the test automation metrics, for example automated versus manual test preparation effort, automated versus manual test execution, or test scripts planned versus executed.

13.1 TEST AUTOMATION METRICS

This section lists a set of metrics used for test automation. The metrics below need to be customised for the organisation's need. It is important to consider more than one item to create suitable metrics and KPIs.

- Test preparation effort, test execution effort metrics and test result analysis effort in test automation

- Metrics related to test script execution, passed, failed, not executed, blocked, deferred and out of scope

- Test execution metrics

- Number of times test script executed: application, environment, release, build

- Test script execution forecast metrics

- Test efficiency metrics

- Automation progress metrics
- Test execution duration metrics
- Percentage of tests passed metrics
- Percentage of tests failed metrics
- Broken build metrics
- Environment downtime metrics
- Test team metrics such as scripts created, defects raised or number of high priority defects per individual
- Automated test effectiveness metrics
- Percentage of tests that were automated
- Test economics metrics:
  - ROI of testing
  - Cost of automation
  - Cost per bug fix, retest and so on
  - Budget variance

13.2 REQUIREMENT AND COVERAGE METRICS

The requirement and test coverage metrics communicate to the stakeholders about what is covered during testing. The examples below of requirements and coverage metrics can be customised in line with the organisation's need.

- User story or requirement coverage metrics
- Acceptance criteria coverage metrics
- Test coverage metrics
- Unit test coverage metrics
- Regression test coverage metrics
- Path coverage metrics
- Percentage of total automated test coverage metrics
- Automated test coverage percentage of manual testing metrics
- Automatable test metrics

13.3 DEFECT METRICS

The metrics provide a measurement of the number and nature of defects found in testing such as defects by priority, defects by severity or defect leakage. The examples below of defect metrics can be customised in line with the organisation's need.

- Defect metrics such as arrival rate or fix rate
- Defect density metrics
- Defect trends metrics
- Defect found in testing metrics: application, environment, stages, release, build
- Defect by severity metrics
- Defect by priority metrics
- Valid/invalid defect metrics
- Defect per team metrics
- Defect per individual metrics
- Defect detection-related metrics
- Defect ageing metrics
- Defect fix rate metrics
- Defect retest metrics
- Defect root cause analysis

13.4 MANAGEMENT REPORTING

Test automation metrics and KPIs based on performance data and analysis are key inputs for test leaders and managers to make decisions and advise other senior stakeholders. Metrics are intended to assist the stakeholders in making more accurate, data-driven decisions. It is essential to know how to prepare and present the test metrics to the stakeholders. In this section, a few examples of management-style metrics for stakeholders are presented.

13.4.1 Product automation metrics

Product automation metrics help to understand and measure the level of automation in different products. The metrics measure the level of automation generally achieved in different testing stages or types such as system testing, regression testing, smoke testing or build validation. Figure 13.1 is an example of product automation metrics.

13.4.2 Defect trends

This metric is a key report that provides visibility of the various defect trends, for example defect arrival and defect closure over time. Figure 13.2 is an example of a defect trend metric.

13.4.3 Test automation trend based on coverage and cost

An automated test coverage metric helps to measure and compare the test automation coverage against manual testing or the cost saving. This is a good indication for test automation effort saving. Figure 13.3 provides an example of the comparison of test automation with manual testing with respect to cost and coverage.

Figure 13.1 Product automation metrics

Product Name	in FY Calender?	Manual Tests				Automated scripts				Client Side	
		SmokeTest	ReleaseTest	Regression Test	TOTAL Tests	SmokeTest	ReleaseTest	RegressionTest	TOTAL Tests	%automated	%manual
Product 1		200	70	349	619	190	50	335	575	93%	7%
Product 2		0	0	328	328	0	0	170	170	52%	48%
Product 3		0	0	260	260	0	0	160	160	62%	38%
Product 4	Yes	50	30	500	580	0	0	312	312	54%	46%
Product 5	Yes	147	0	1000	1147	142	0	850	992	86%	14%
Product 6	Yes	50	0	300	350	50	0	0	50	14%	86%
Product 7	Yes	0	0	5577	5577	42	0	2050	2092	38%	62%
Product 8	Yes	0	0	2950	2950	0	0	41	41	1%	99%
Product 9	Yes			10000	10000	141	0	607	748	7%	93%
Product 10	Yes	0	0	3071	3071	0	0	378	378	12%	88%
Product 11	Yes	0	0	127	127	0	0	51	51	40%	60%
Product 12	Yes	0	0	350	350	0	0	250	250	71%	29%

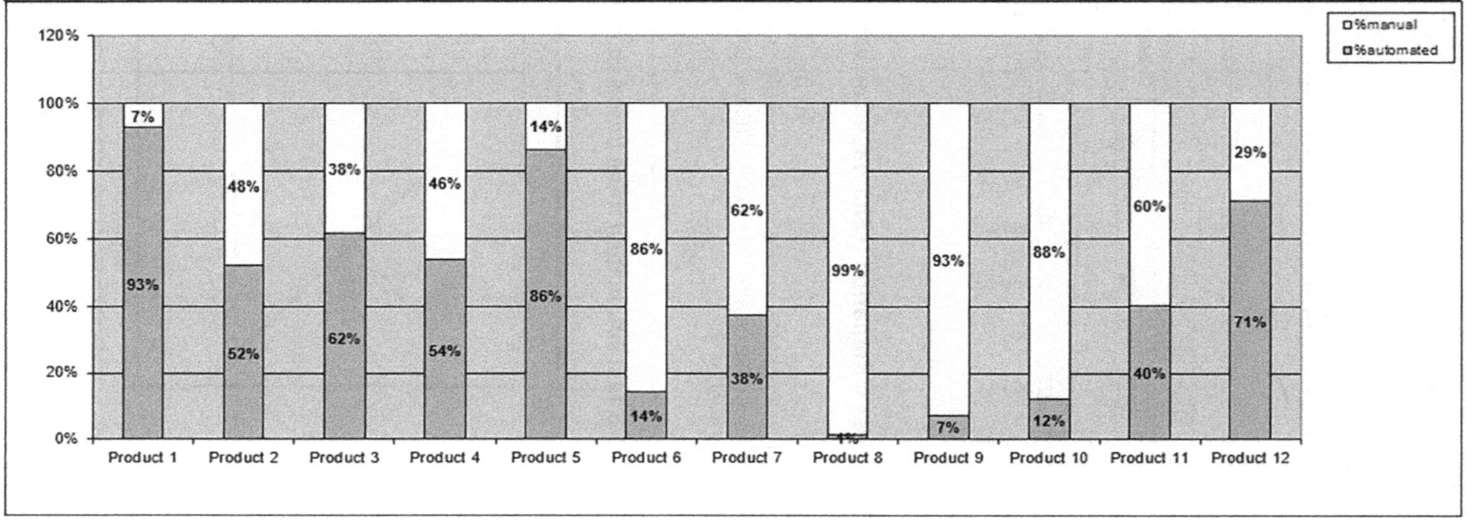

Chart (□ %manual, ■ %automated):

Product	%manual	%automated
Product 1	7%	93%
Product 2	48%	52%
Product 3	38%	62%
Product 4	46%	54%
Product 5	14%	86%
Product 6	86%	14%
Product 7	62%	38%
Product 8	99%	1%
Product 9	93%	7%
Product 10	88%	12%
Product 11	60%	40%
Product 12	29%	71%

Figure 13.2 Defect trends

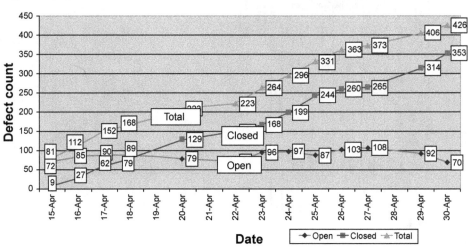

13.4.4 Test automation trend based on effort and quality

Effort saving is generally a key priority for the IT leadership and metrics related to effort and quality are some of the most valuable sources of information for management. Figure 13.4 is an example comparing test automation with manual testing in respect of effort and quality.

Figure 13.3 Test automation trend based on coverage and cost

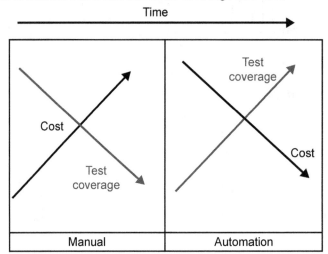

Figure 13.4 Test automation trend based on effort and quality

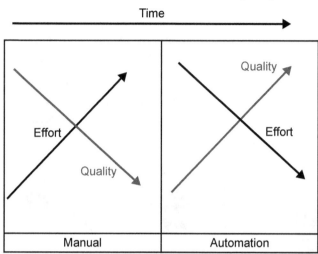

13.4.5 Test automation trend based on test execution rate

Figure 13.5 is an example of the comparison of test automation with manual testing with respect to test execution. This is a key indication for testing achieved by test automation.

Figure 13.5 Test automation trend based on test execution rate

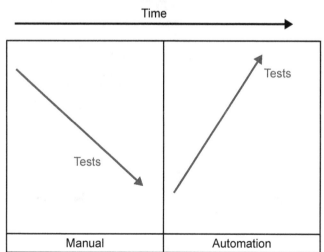

13.4.6 Test automation defect removal efficiency

Test automation defect removal efficiency provides a measurement of the removal of defects from the product or release. It is calculated as a ratio of defects found in test automation to total number of defects found across manual and automated testing. Figure 13.6 shows an example of test automation defect removal efficiency metrics.

Figure 13.6 Test automation defect removal efficiency

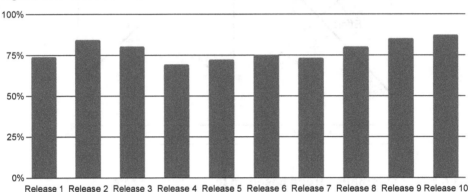

13.4.7 Test automation defect root cause analysis

Root cause analysis is used in software testing to identify the cause of the defects. The metrics related to root cause analysis helps management to identify the key areas for quality improvement. Figure 13.7 is an example of test automation defect root cause analysis.

Figure 13.7 Test automation defect root cause analysis

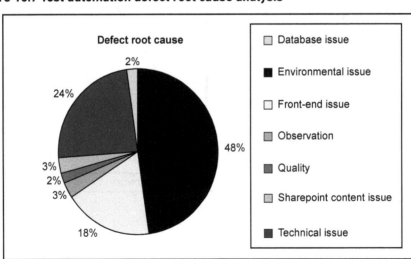

13.4.8 Test automation defect density

The defect density helps to identify the most problematic area of a product. This will lead to further root cause analysis and helps in understanding the efficiency of testing. Figure 13.8 is an example of test automation defect density metrics.

13.4.9 Test automation environment downtime metrics

The test automation environment downtime-related metrics helps in understanding the impact of test environments on test automation. This helps management in the decision-making process, for example by allocating a dedicated environment for test automation to avoid sharing environment with other teams and avoid interruptions. Figure 13.9 is an example of test automation environment downtime metrics.

13.5 AUTOMATED TEST METRICS

Test automation metrics allow the results of automated testing to be measured, and automated test metrics create metrics automatically without human intervention, for example automated dashboards for metrics.

Figure 13.8 Test automation defect density

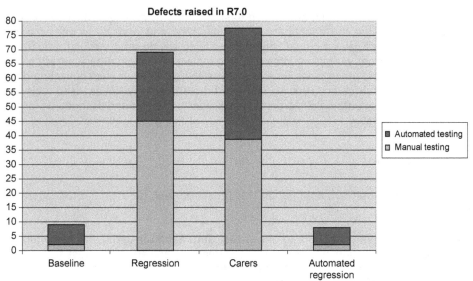

Figure 13.9 Test automation environment downtime metrics

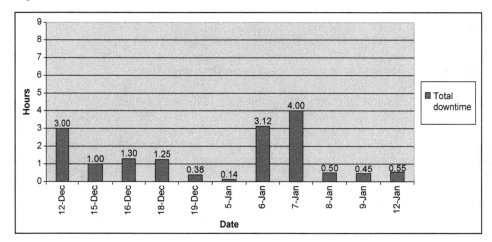

Test metrics automation is the process through which the reports, metrics and dashboards are created and automatically updated using software. The gathered data, metrics, charts and so on can be delivered to specific email addresses on a regular basis with automatic email dispatches.

Test metrics automation and automated dashboards provide a centralised reporting platform. This helps the dashboard to display KPIs and metrics in one location. Automation reports have multiple benefits, including saving time, viewing robust and up-to-date data from various sources in a single place, enhancing the accuracy of the information and providing more time to analyse the data rather than collecting data. Automating the metrics and reports will save time and energy. Widely used tools like MS Power BI help to automate and personalise reports, metrics and dashboards for the organisation's needs.

13.6 SUMMARY

This chapter summarised automation test measurement, metrics and management presentation.

The purpose of this chapter is to assist test automation leaders and managers to measure test automation in the organisation effectively and efficiently. Test automation metrics are measurements of the implemented automated testing process. This includes past, present and future metrics.

14 CONCLUSION

The recent downfalls of Thomas Cook, Maplin, Toys R Us, HMV, BHS and Debenhams are largely due to the delay of digitalisation or technological innovations. On the other hand, the more technologically forward-thinking companies such as Amazon, eBay, Airbnb, Uber, IKEA, Deliveroo and Greggs used IT technology to enhance their growth for their business model and sustainability. Being the first or quick in the market is one of the key factors for the success of many businesses. In recent years, organisations have been facing enormous challenges to reduce the turnaround time to be the first in the market with new products, and IT organisations face enormous challenges to shrink the software development timeframe. Many software development companies use Agile and DevOps primarily to reduce the timeframe. Test automation is significant and is the core of the DevOps and Agile way of software development in reducing the testing effort and supporting frequent releases and regression testing.

Today, test automation is considered one of the most effective ways to enhance the coverage, consistency and effectiveness of software testing. It is heavily used to improve test efficiency during implementation, post-live or BAU stages. Although test automation has redefined the way testing is viewed, it is still difficult to implement automation as it requires high investment, time, long-term commitment from the organisation, continuous support and maintenance to produce the desired outcome. Test automation returns many benefits such as cost reduction, accuracy and the expected results in the long run if understood and implemented correctly.

This is a book about test automation and automated testing, which is part of a software testing discipline that has evolved over the last few decades, and has the potential to offer great benefit to IT organisations by ensuring that there is alignment between business needs. This book provides guidance to IT leadership on test automation, reflecting the scope and methods used. Many organisations introduce test automation as an IT innovation; however, there persists a lack of understanding on the needs for and benefits of test automation and this often creates more questions than answers.

This book has endeavoured to address the following:

- 'Whys' and 'Whats' of test automation
- 'Hows' and 'Whens' of test automation
- Skills required for test automation
- How test automation adds value to the organisation

I would like to wish you the very best in your test automation journey!

APPENDICES

APPENDIX A: CODING STANDARDS AND COMMENTS

CODING STANDARDS

Coding standards and good practices are essential for software development and scripting, irrespective of the programming languages and coding method followed. The purpose of coding standards is to write clean code, and the key principles of coding standards are:

1. Keep the code simple.

2. Don't repeat yourself (DRY). This is aimed at avoiding repetition of information.

3. Be consistent when applying coding standards.

4. Comment the code for review and maintenance.

The key benefits of using coding standards are:

1. It helps compliance with industry standards.

2. It maintains code quality.

3. It reduces development costs and accelerates time to market.

CODING COMMENTS

Comments are used to help anyone who is attempting to understand and maintain the actual code. For consistency, comments should adhere to the following guidelines:

- Comments should be included in scripts to provide an explanation of what the code is doing, and to make it easier for the reader to understand and maintain it. For this reason, it is not necessary to try to comment each individual line or to describe obvious code, but it is necessary to comment major blocks of code.

- Comments lines should begin with the ' character indented at the same level as the code they are documenting.

- Code comments should be written in clear, concise English with correct spelling and grammar.

- Comments should not be used to teach the reader how to program.

- The purpose of comments is to increase your code readability. Often using good variable and method names makes the code easier to read than if too many comments are present.

- When 'commenting out' code, include a description of why you did so. If the code is not necessary, it should be deleted.

- Do not use 'clutter' comments, such as an entire line of asterisks. Instead, use a single line space to separate comments from code.

APPENDIX B: SAMPLE TEST AUTOMATION FRAMEWORK

Figure A.1 shows a sample framework created in Selenium and C# programming language for an ecommerce application.

Figure A.1 Test automation framework folder structure

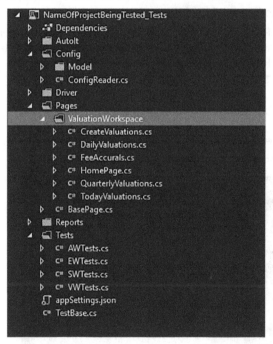

The key folder details are provided in Table A.1 for the framework.

Table A.1 Test automation framework folder description

Folders	Description
Config	Stores the information that remains static throughout the framework such as browser-specific information, application URL, screenshots path.
Extensions	Extension Methods (C# Programming Guide): 'Extension methods enable you to "add" methods to existing types without creating a new derived type, recompiling or otherwise modifying the original type. Extension methods are static methods, but they are called as if they were instance methods on the extended type' (Microsoft, 2020).
Factory	Driver Factory: ThreadSafe Driver for running parallel test execution.
Browser Factory	This allows a browser instance to launch based on the parameter, and creates the webdriver object for the given browser.
Helpers	Helper class is used for Date, FileType, Name and WaitHelper.
Logger	This module will log messages to test output.
Reports	Extent Reports has been used in this case. It generates HTML reports and maintains logs to include the screenshots of failed test cases in the Extent Report.

Figure A.2 provides the framework shown in Figure A.1 used for a real-life project.

Figure A.2 Test automation framework for a real-life project

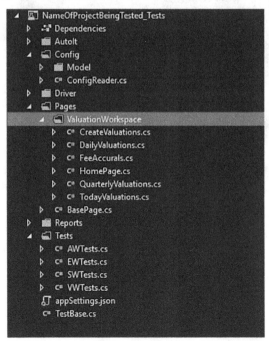

The folder details are provided in Table A.2 for the real-life project framework.

Table A.2 Test automation framework folder description for a real-life project

Folders	Description
AutoIt	AutoIt v3 is a freeware like BASIC scripting language designed for automating the Windows GUI and general scripting. It uses a combination of simulated keystrokes, mouse movement and window/control manipulation in order to automate tasks, which are not possible with Selenium.
Config	Configuration instance to get the values from the appSettings.json file.
Model	This is the model class for the appSettings file.
Pages	Holds the page object for the application. The framework creates a new folder for each project. Apart from Homepage all other pages will inherit from BasePage.
Reports	Extent reports are stored in the Reports folder. In this case it overwrites the file in each run.
Tests	Stores all the tests.

In this framework, the test cases were executed in a parallel run. The parallel execution helps to save time in test execution; for example in a normal scenario, if it requires 100 minutes to run 100 tests, in parallel execution if five threads are used at the same time, it takes only 20 minutes to run 100 tests (100/5 = 20) (Figures A.3 and A.4).

Figure A.3 Test automation framework test result 1

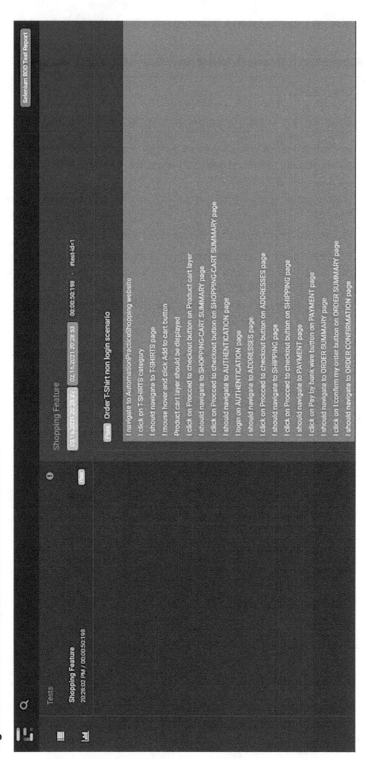

Figure A.4 Test automation framework test result 2

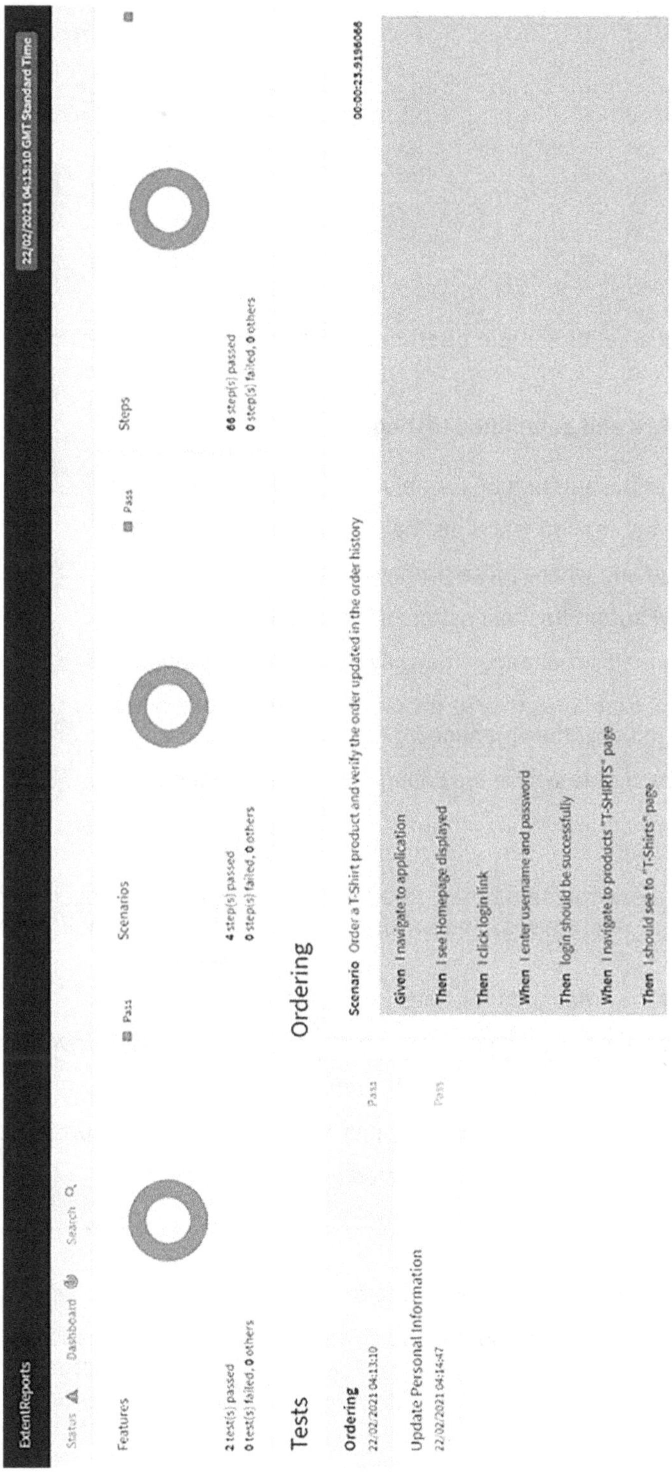

ExtentReports

22/02/2021 04:13:10 GMT Standard Time

Status Dashboard Search

Features

Pass Scenarios

Pass Steps

2 test(s) passed
0 test(s) failed, 0 others

4 step(s) passed
0 step(s) failed, 0 others

66 step(s) passed
0 step(s) failed, 0 others

Tests

Ordering
22/02/2021 04:13:10   Pass

Update Personal Information
22/02/2021 04:14:47   Pass

Ordering

Scenario Order a T-Shirt product and verify the order updated in the order history   00:00:23.9198066

Given I navigate to application

Then I see Homepage displayed

Then I click login link

When I enter username and password

Then login should be successfully

When I navigate to products "T-SHIRTS" page

Then I should see to "T-Shirts" page

APPENDIX C: SAMPLE INTERVIEW QUESTIONS

This section covers a set of sample questions to ask the candidate during the hiring process.

Test tools, framework and automated testing:

- Ask them about the automation framework they have worked with.
- Talk about the automated test tools that they have worked with.
- Ask them when and where these frameworks and tools are used.
- Talk about how automated test scripts are derived.
- Talk about the major challenges they have faced in automation.
- Ask them about a situation they've encountered where the tool was not compatible with the SUT and how they managed.
- Ask them about a case where they doubted their automation skills.

Test environments:

- Ask them what environments they have tested in, that is, web/webserver, client–server, desktop, on-premises, cloud, hybrid, secure, crypto, Unix, Linux, mainframe and so on.
- Ask them how they managed releases in different environments.
- Ask them how they developed scripts or frameworks for different environments.
- Ask them how they developed scripts for multiple browsers.
- Ask them what they did differently for Agile, Waterfall and DevOps projects.
- Ask them what they did for fully integrated environments.

Managing testing delivery:

- Ask them what documents are produced for automated testing and why.
- Ask them to talk about test policy, test strategies, test plans, test approach, test reporting, test automation metrics and the purposes of these documents.

- Get them to talk about automation test architecture, automated test scripts, what the contents of these are, what the purpose of these documents is, where they fit into the testing life cycle, who writes them, and so on.

- Ask them what stages they have been involved in for automated testing.

- Ask them about the automation team in DevOps, Waterfall and Agile.

- Ask them about their current team, team structure, role and typical day in the team.

Product or project scenario:

- There are two weeks until the milestone is due, and development have delivered late, leaving four weeks' worth of work left to do. The milestone cannot be moved and is still due in two weeks' time. What do you do for testing? How does your automated test suite help?

- The expectation from the team is to complete automation within the Sprint itself. What is your approach in this case?

- The project requirements and SUT change frequently or the requirement is to automate a dynamically changing UI. What is your approach for test automation and why?

- Automate a solution not compatible with the tool. What is your approach for test automation and why?

- There is a two week Sprint and the development team have not completed any features after week 1. What is your approach for automation? What do you do in week 1?

Talking about themselves:

- Ask them what qualities they think makes a good automation subject matter expert and get them to relate the quality to themselves, giving examples from their past work assignments to illustrate this.

- Ask them about how good they are at working under pressure.

- Ask them about being self-motivated.

- Ask them for examples, basing them on experiences from previous work assignments.

Testing process:

- Get them to describe the test automation life cycle and stages. Ask them to talk about testing methods and models. They should include things like the V-model, continuous integration, DevOps, Waterfall model, iterative and Agile/Scrum and so on.

- Delve into the test automation applicable to previous testing assignments that they have worked on.

- Ask them to describe the automated test scripting process, how they create scripts, what they are based on, who gets involved during the writing and reviewing stages, and how the quality of the writing is maintained and so on.

- Talk about what constitutes an automation suite and how the test execution phase is conducted and so on.

Career aspiration:

- Ask them why they like testing and test automation.

- Ask them what they do not like in test automation.

- Ask them what their medium-to-long term plans are. What are their long-term goals?

- Ask them what they can offer as a member of the testing team. Why are they the best candidate for the role?

APPENDIX D: SAMPLE SKILL SET OF TEST AUTOMATION ENGINEERS

This section provides a few examples of sample skill sets of test automation engineers.

TEST AUTOMATION ENGINEER: FUNCTIONAL TESTING

Summary:

- Sound experience in IT industry including software development, software testing and test automation.
- Experience in leading a test team of 3–5 testers (automation and manual).
- Proficient in writing and executing test scripts with QTP, Selenium, Coded UI and Quality Center, extensive use of test management tools such as Quality Centre and Jira.
- Proficient in designing automation framework (hybrid framework supported by keyword-driven) and automation test suite.
- Strong experience working in Agile (Scrum) environment and active involvement in release planning, user-story workshops, Sprint planning, iteration, stand-up, demo, review and retrospective as a key member from testing.
- Ability to communicate effectively with cross-functional teams and members such as developers, business analyst, product owner, Scrum master and other Scrum teams.
- Develop test automation framework using Selenium, Webdriver with C#, NUnit, SpecFlow, BDD and Page Object Model.
- Design and build test automation framework in Coded UI (C#) and Selenium.

Responsibilities:

- Develop automated test cases using Coded UI, Page Objects pattern and BDD (SpecFlow).
- Develop automation scripts for web service (API) testing using C#, and SOAP UI.
- Deploy and maintain system test environment (create PowerShell scripts).
- Train project manager, developers, business analysts and testers in using Microsoft Azure DevOps.

- Develop automated test cases using Page Objects pattern and BDD (SpecFlow) using C# webdriver.
- Service-oriented architecture (SOA) testing for payments APIs using SOAP UI and C# framework for web services testing.
- Integrate C# webdriver test framework with the continuous application deployment in TeamCity so the build target will automatically run automation scripts after build deployment.

Technical skills:

- **Automated test tools.** Selenium RC and WebDriver, Coded UI, HP Quick Test Professional, WinRunner, Maven, TestNG, Eclipse and .Net.
- **Test management tools.** Quality Center, Bugzilla, Jira, PVCS Tracker, .Net, MS SharePoint, Microsoft Test Manager and Version one.
- **Programming.** C# and Java.
- **Version control.** Visual Studio Team Edition, PVCS Version Manager, Microsoft Visual Source Safe, Team Foundation Server and Microsoft Azure DevOps.
- **Database.** MS Access, MS SQL Server, Oracle.

TEST AUTOMATION ENGINEER: PERFORMANCE TESTING

Summary:

- Performance test lead coordinating a team of 10 people
- LoadRunner for 10 years using a variety of different protocols
- ISTQB Advanced Test Manager qualification

Technical skills:

- LoadRunner
- HTML/HTTP
- Web Services/SOAP
- WinSock
- Tuxedo
- Web Service, XML and VB.Net
- Oracle: PL/SQL and SQL
- UNIX shell scripting
- Windows scripting (WMI, WSH and batch scripts)
- SharePoint development

APPENDIX E: TEMPLATES

This appendix provides sample templates and structures, which need to be customised for your organisation. Each section of the table of contents needs to be expanded based on your specific requirements and may be different for your organisation.

TEMPLATE 1: SAMPLE TEST POLICY

Table of contents

1. Purpose
2. Scope
3. Definitions
4. Policy
    - 4.1 Risk-based testing
    - 4.2 Test artefacts
    - 4.3 Automated testing
    - 4.4 Maintenance
    - 4.5 Regression testing
    - 4.6 Requirements (all types)
    - 4.7 Test early
    - 4.8 Non-functional testing
    - 4.9 Test deliverables
    - 4.10 Resource skill sets
    - 4.11 Reporting
    - 4.12 Test metrics, KPIs and measures
    - 4.13 Test traceability
    - 4.14 Test governance
    - 4.15 Lessons learnt
    - 4.16 Software testing standards compliance

1. Purpose

This policy is to ensure the following:

- Test activities are performed with governance and control, and comply with defined test processes.
- Test approaches are consistent and measurable.
- Risk-based testing is utilised on all projects.
- Where applicable, automated testing is considered, implemented and maintained to maximise test efficiency and cost savings.
- A proactive test approach is taken to ensure maximum productivity and to lower test activity costs.

2. Scope

This test policy will apply to all IT led programmes.

3. Definitions

IT: information technology

IT stakeholders: accountable domain leads, senior project leaders, project leaders, domain senior test managers, and test managers or leaders within or representing IT and business

4. Policy

Test policy is based on industry best practices such as ISTQB and PRINCE2.

- Risk-based testing – Risk shall be set during a number of workshops with all key stakeholders to ascertain the test risk of the requirements, features and functions within the project. Requirement risk analysis can still be conducted if the project deems it necessary, and this will be performed using industry standard best practices within the test management tool.
- Test artefacts – All test artefacts such as test plans, test scripts, frameworks, test data and defects will be created and maintained throughout the life cycle.
- Automated testing – All IT programmes and products should complete an automated testing feasibility study using the predefined automated testing feasibility template or tool. The tool contains a set of questions around the software solution and the long-term impact.
- Maintenance – All automated test suites or packages must be maintained; the test manager is responsible for ensuring that the automated testing is handed over to service run/BAU team and maintained for future use.

- Regression testing – All regression testing is expected to be automated unless the automation suitability checklist identifies that there is insufficient return on investment (ROI).

- Requirements – Before testing activities start the requirements must be signed off. Once the requirements are in the test management tool and tests are linked to the requirements, then all further requirement updates must be performed within the test management tool. This ensures all changes are tracked and the audit trail is in place. Requirement test coverage of 100 per cent will be achieved unless there is prior agreement with the project stakeholders due to descoped requirements, in which case the requirements will be marked accordingly but will remain within the test management tool.

- Test early – Early life cycle testing must be adopted.

- Non-functional testing – Non-functional testing should include performance, load, stress, capacity, IT service continuity (ITSC) and security testing. The results of all non-functional testing must be stored within the test management tool, to demonstrate full coverage, risk and direct cover status.

- Test deliverables – All test deliverables must be available at any time to be reviewed and audited.

- Resource skill sets – All resources involved in any aspect of testing must have sufficient test skills to fulfil the role. They should have good domain knowledge and understanding of the business processes of the system under test.

- Reporting – Reporting will occur during the complete life cycle to ensure that there is a clear prediction of completion date, coverage and risks. All status reports must be generated from the test management tool. The reports are to include requirement coverage, direct cover status, defect logging, test write progress, and test execution status against releases and cycles.

- Test metrics, KPIs and measures – Project level performance metrics and measures will be captured throughout the project.

- Test traceability – All test activities must be included in the overall project plan and should be accessible.

- Test governance – All programmes must have a single point of accountability for the overall delivery of all test activities on the project to ensure consistency and that testing is efficient and cost-effective.

- Lessons learnt – Upon the completion of a project, project teams should take part in a test-focused lessons learnt session to establish where test efficiencies can be improved, costs reduced and risks mitigated for future projects.

- Software testing standards compliance – All testing activities should comply with ISO/IEC/IEEE 29119 software testing standards and those outlined in the test policy document. http://softwaretestingstandard.org

- Test tools – This section lists tool selection.

- Exceptions – Any exception must be made with deliberate consideration of all the associated risks, and exceptions must be agreed with the stakeholders.

TEMPLATE 2: SAMPLE TEST AUTOMATION PLAN

Table of contents

1. Document control

1.1 Document information

1.2 Version history

This document is subject to change control after it has been formally signed off. All subsequent changes to this document must adhere to the document control process and change management process.

1.3 Distribution list

1.4 Approvals

1.5 Source documents

The inputs for this document have been based on, or should be read in conjunction with, the following documents:

- Document 1 (example)
- Document 2 (example)
- Document 3 (example)

2. Introduction

This document is the automation test plan for the project. This test plan includes the objective of test automation, which is to determine whether the functionality of the system under test is as per the requirements. The objective of automation is to automate test scenarios and test cases that are common functionalities across various modules undergoing the project implementation. The same test suite can be executed to perform a regression test run without investing time and effort. The intended audience of the automation test plan document includes the project manager, QA manager and entire functional testing team and development/solution teams.

3. Test scope

3.1 Activities

Area	In scope and out of scope
Feasibility and planning	Analyse the functional test cases for automation regression testing
	Identify the test cases for automation; prepare the automation plan
	Identify the repetitive tasks and create the test automation framework
	Identify relatively stable areas of the application over volatile ones for automation
	The common workflows to be performed by the end-user will be automated first
Test case automation	System under test: [Name]
	Web application will be automated using [Tool name]
	Automation feasible test case count: [Number]
	Tool: [Tool name] with [Programming language name] as programming language
	Framework: Hybrid-based framework will be used for automation
	Language versions: Application version – English
	Delivery plan: The automation will be delivered with the below number of test cases that will automate
Business scope	Areas of application that have least priority for the business will be out of scope
Test automation	Areas of application for which there is no functional or technical reference present in the requirements document

3.2 Entry and exit criteria

Entry criteria:

- Functional test plan and test cases – reviewed and signed off
- Application build with proper release notes from the development team
- Test data required for the test
- Environment setup (e.g. UFT installation and integration with ALM)
- All the available system set up (OS machine and devices)
- Application knowledge transition

Exit criteria:

- All the feasible test cases have been automated.
- The customer has signed off on the automated test cases and automated testing scope.

3.3 Deliverables

- Automation test plan
- Automation test scripts
- Automation test execution summary report

3.4 Test approach

The focus of the test automation is to automate the steps used for testing the application manually. It will be designed based on the functional test scenario identified and thus verify the crucial functional units. The test automation team will utilise a data-driven approach to test, analyse and validate the application.

The test automation consists of the following high-level activities:

- Identify high-level test scenarios.
- Identify the reusable components.
- Create a design document for the identified reusable components and specific scripts.
- Set up the data for test automation.
- Develop automation test scripts.
- Review automation test scripts.
- Execute automation test runs through test management tool.
- Carry out the defect analysis during the automation test execution phase.
- Report/track automation test defects.
- Report test summary results.

3.5 Governance

The QA responsibilities assigned are listed as follows:

- Create automation test plan.
- Estimation, planning and control.
- Obtain the sign off for automation test plan.
- Manual test scenarios and test scripts review.
- Automation test scenario and script review.
- Oversee test execution and defect management.
- Obtain the sign-off for test summary report.

4. Test automation

Test automation will involve automating the functional test scenario identified. The main objective of test automation is to validate the functionality of the application.

4.1 Testing techniques

The test automation team will utilise a data-driven approach to test, analyse and validate the SUT. The data-driven approach is applied for the test automation activity as it separates data from the automation scripts. This allows the features to be tested with a different set of input values and also by parameterisation of scripts, and scripts can be run on multiple data sets to verify that the application works as expected. It allows process flows to be tested first and then the end-to-end scenarios interacting with external systems will be tested. In future any change in data requires only the data set to be updated and not the automated script.

4.2 Testing tools (example)

Test process	Tools
Test automation	• HP Unified Function Testing
Test management	• HP ALM
Defect management	• HP ALM
Test execution (scenarios/scripts)	• HP ALM and Unified Functional Testing (UFT)
Test data creation	• HP UFT inbuilt MS Excel spreadsheet (data table)

4.3 Suspension and resumption criteria

Testing can be suspended under the following circumstances:

- Connection issues with the automation tool licence server.
- Non-availability of system under test.
- Network fluctuations.
- Unresolved build errors.
- Unscheduled technical disruptions to the testing environments.

If a test case fails, testing will be suspended for all dependent features. The failed test case will be logged in a defect log along with a description of the failure.

Testing will resume, from the point at which it was suspended, when the items listed above are resolved.

If a full suspension occurs and can be pinpointed to a specific software release, then the re-deployment of the previous successful software release will be considered if further testing of that and earlier increments is thought to be of benefit.

5. Test automation framework approach

5.1 Approach

The approach section provides the solution test automation. At high level, the test automation life cycles are based on following industry standards.

The SUT automation will be carried out based on the above defined automation approach, methodology and the following framework. This automation approach and framework advocates the usage of data-driven testing within the industry standard methodology.

The framework consists of the following elements:

- **Data-driven scripts**
- **Business components**
- **Application area**
  - Object repository
  - Utility and generic functions
  - Reporting functions
- **Automated test execution**
  - Individual and batch execution
- **Report manager**
  - Excel reports
  - UFT standard reports

5.2 Test data

The test data will be created according to the parameterisations done during the process of automation test script creation and will be maintained in the MS Excel spreadsheet. A separate MS Excel sheet will be maintained for each test script, it will contain the data set used by the script and will be stored in the test management tool. This will ease any changes that need to be made to the data set.

5.3 Execution flow

- All the libraries will be associated with runtime (loaded at runtime).
- The test automation script will read the business component tests mentioned in the process flow tab and execute them one by one.
- All the components mentioned in business library will be executed by the automation script.
- The report will be generated in MS Excel and HTML formats.

5.4 Test preparation

The following test preparation activities are required:

- Verify connectivity between the automation tool and licence server.
- Confirm test management tool and automation tool connectivity for storing the test script and execution.

5.5 Test environment

Testing is truly enabled by a system environment plan. The following points outline the code promotion path for the project:

- The automation team requires access to all the required tools.
- Test environment: functional and end-to-end (E2E).
- Database access to prepare the test data.

5.6 Automation test execution

Automation script execution will be carried out through the test management tool. Defects found during each cycle will be reported through the defect tracker. The team will record any defects using the defect management process defined. Automation test scripts will be re-executed to ensure that the bug fixes have not affected any other part of the application. As test execution progresses, the test governance lead will constantly update and modify the test execution matrix according to which test scripts have passed and failed, which cannot be run due to dependencies on other things, and which need to be retested because of failure or defects.

A test summary report will be created that will contain information on the total number of test cases executed and defects found during each test cycle.

5.7 Automation test management

The following test management activities have been defined:

- Manage and control automation test life cycle.
- Manage the automation test scripts.
- Perform automation test execution.
- Investigate and resolve discrepancies.
- Prepare test results and defect reports.
- Review test results and defect lists.
- Create test summary report.
- Validate whether or not all testing tasks have been completed.
- Obtain sign-off.

5.8 Automation test schedule

The following are examples of a test automation schedule in two different formats.

Activity	Efforts (in person days)	Start date	End date
Test plan/Test strategy preparation	10		
System study/Knowledge transfer Query/Access	5		
Test data preparation	1		
Effort required generating automation framework	7		
Script development + UFT basic framework as per test flow	12		
Integration and dry run	0.88		
Reviews and incorporating review comments	0.55		
Project management effort	3.2		
Contingency	1.6		
Customer demonstration	1		
Total effort [in person days]	42.23		
Total person months required	1.91		

6. Test metrics

Accurate reporting of test results is just as critical as the actual testing itself. Effective test metrics provide test management ability to report progress of the test against plan in several ways. A baseline level of metrics should indicate planned and actual test execution results, as well as all active defects by status. More sophisticated metrics provide trend analysis and some outcome projection capabilities. The test team will create and communicate the metrics periodically.

7. Risks, assumptions, issues and dependencies (RAID)

8. Responsible, accountable, consulted and informed (RACI) matrix

TEMPLATE 3: COST-BENEFIT ANALYSIS

Factors to be considered:

- Total cost of the test automation versus 100 per cent manual testing.
- When will the full ROI be realised?
- What are potential risks or challenges?

Figure A.5 Tentative high-level plan and schedule

Weeks					-22	-21	-20	-19	-18	-17	-16	-15	-14	-13	-12	-11	-10	-9	-8	-7	-6	-5	-4	-3	-2	-1	0
Tasks	RESOURCE	START (W/C)	END (W/E)		28-Jun	5-Jul	12-Jul	19-Jul	26-Jul	2-Aug	9-Aug	16-Aug	23-Aug	30-Aug	6-Sep	13-Sep	20-Sep	27-Sep	4-Oct	11-Oct	18-Oct	25-Oct	1-Nov	8-Nov	15-Nov	22-Nov	29-Nov
Test Strategy				4 weeks																							
Unit testing				9 weeks																							
Test Automation plan preparation				4 weeks																							
SUT Study /KT/Query/Access				2 week																							
Test Data Preparation				4 week																							
Test Automation framework development				4 weeks																							
Test Automation script development				8 weeks																							
Integration and dry run				6 weeks																							
Reviews				4 weeks																							
Customer Demonstration				2 weeks																							

Cost-benefit analysis		
What to include in a cost-benefit analysis report	**Value (positive or negative)**	**Additional comments**
Requirements and technology related		
What is the SUT architecture and technology?		
How much can be automated (% of test cases or features that can possibly be automated)?		
What is the level of test automation in test design, test reporting and test execution?		
Cost of test automation		
What is the estimated cost if automation is used in this product/project? (Total cost [£ = hardware + software + development cost + execution cost + maintenance cost])		
Tools including software licences and hardware capacity		
Scripts creation effort		
Scripts or framework maintenance effort		
Execution effort		
What is the estimated cost if 100% manual testing is used? (Total cost [£ = hardware + software + development cost + execution cost + maintenance cost])		
Training cost		
Resources related		
Tools (software and hardware) and their cost		
People		
Test environments and data related		
Environments		
Test data		
Other technologies		
Has the prototype been considered?		
Has a cloud-based test environment been considered?		
Has the simulator/emulator/virtualisation been considered?		

Cost-benefit analysis		
What to include in a cost-benefit analysis report	**Value (positive or negative)**	**Additional comments**
Transition plan		
BAU/post-live support		
Tools support		
Skills		
Risks relating to the whole automation delivery		
Risk		
Tangible benefits		
Benefit 1		
Intangible benefits		
Benefit 1		

TEMPLATE 4: AUTOMATION SUITABILITY CHECKLIST

Automation suitability checklist				
Area	**Consideration**	**Weighting (%)**	**Assessment**	**Comments**
Project	Duration of project (months)	15% (example)		
	Required execution frequency			
	Complexity			
	Stability			
	BAU/post-live support (months)			
	BAU execution frequency			
Ability to execute	Test data			
	Environment			
	Prerequisites			
Test quality	Requirements stable and agreed			
	Test cases available			
	Risk analysis of existing tests			
	GUI or backend testing			
Handover	Team available for handover			

Key
< 50%: Not ideally suited to automation
50–60%: Automation is viable
60–75%: Suitable for automation
> 75%: Great automation candidate

Area summary	Value
Project %	
Ability to execute %	
Test quality %	
Handover %	
Total %	

Project suitability for automation	Yes/No

TEMPLATE 5: TOOL EVALUATION

Tool evaluation	
Tool name	[Name of the tool]

Vendor/ community	Vendor/community support	
	Vendor size	
	Vendor support available	
	Vendor refund policy	
	Any initial training by vendor	
Costs	Is this a licenced or open-source tool?	
	Purchase costs	
	Support costs	
	Maintenance costs	
	Type of licences	
Skills	Are the staff skilled in tool usage?	
	Does the tool require development knowledge?	
	What coding language knowledge is required?	

Project use	Does the tool support the technology stack?	
	Single product use? Or multiple projects?	
Integration	Integration with test management tool	
	Import/export capability	

Tool evaluations	Test automation tool comparison	Tool one	Tool two	Tool three
	Pros			
	Cons			
	Reason for not selecting tool			
	Reason for selecting a tool			

TEMPLATE 6: ROI TREND

Regression testing scope	
Approximate total no. of test cases	414
Automation %	75%
Approximate test cases in automation scope	316
Approximate test execution cycles in year	5 to 6

Testing type	Test design effort		Test execution effort	
	One test case in mins	In scope test cases person days	One test case in mins	In scope test cases person days
Manual	TBC	195	TBC	110
Automation	TBC	444	TBC	43

Total effort for given cycle in hours		
Cycles	Manual	Automation
Dev	1465	3330
1	2292	3738
2	3119	4107
3	3946	4433
4	4773	4759
5	5600	5085
6	6427	5411
7	7254	5737
8	8081	6063
9	8908	6389
10	9735	6715
11	10562	7041
12	11389	7367
13	12216	7693
14	13043	8019
15	13870	8345

Figure A.6 ROI chart for automation

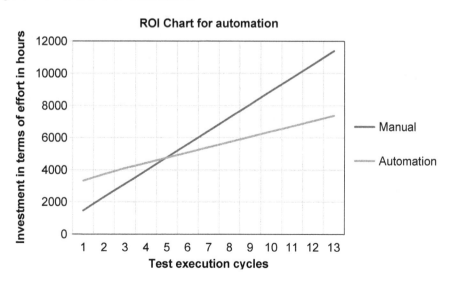

TEMPLATE 7: ROI

Benefits/value estimation:

No.	Test automation value	Details	Individual weightage (high/ medium/low)	Actual value
1	Test coverage			
2	Less resources			
3	Defects			
4	Test automation tangible benefits			
5	Test automation intangible benefits			
6	Test automation – platform support			
7	Scope of the automation			
8	Break-even			
9	Accuracy			
10	Long-term benefits			
11	Reusability and repeatability			
12	Availability			
13	Job satisfaction			
14	Automated scripts execution			
15	Application releases			
16	Performance testing			
17	Repetitive tasks			
18	Test automation value addition			
19	Long-term benefits			
20	Time			
21	Data validation			
22	Skill availability for test automation			
23	Tools availability			

Cost estimation:

No.	Test automation: value	Details	Individual weightage (high/medium/low)	Actual value
1	Framework cost			
2	Test scripts cost			
3	Execution cost			
4	Cost of test automation			
5	Result analysis			
6	Other costs			
7	Manual testing			
8	Long-term costs			
9	Tools			
10	Defects management			
11	Complexity of application			
12	Test scripts maintenance			
13	Tools additional features			
14	Domain knowledge			
15	Test setup and environments			
16	Application changes			
17	Time			
18	Multiple platforms			

Factors:

The questions and factors shown below should be carefully considered for the ROI calculation

No.	Factors	Check
1	How much time do you have for training?	
2	What is the scope for test automation?	
3	Do you want to establish traceability for compliance?	
4	Is it record and playback test automation?	
5	Does the application support automation?	
6	Does the testing tool support automation?	
7	Does the SUT architecture and technology support automation?	

8	Does the application domain support automation?	
9	Does the test case support automation?	
10	What is the test coverage of automation?	
11	Does it catch defects sooner in the software life cycle?	
12	Is the development of scripts a parallel effort to SUT development?	
13	Does it bring value to the customer?	
14	Does the project require automation?	
15	Do you have a test automation strategy?	

APPENDIX F: TEST AUTOMATION – INDUSTRY EXAMPLES

The examples below are real-life examples from the test industry.

Customer	A large retail company in Europe.
Industry	Manufacturing, retail and distribution (MRD).
Project description	A major European retail company had a requirement to review the promotional capabilities to increase the number of concurrent promotions that could be supported, and also increase the flexibility of the promotion offers that could be provided.
Solution	Upgrade the software.
Team	Test team size was four including the test manager/Scrum master.
Testing	Component and component integration testing executed by the third-party supplier, system testing, performance testing, operational acceptance testing, system integration testing and user acceptance testing executed by the SI and the client.
Automation/ tool	A major test management tool used for defect reporting and tracking.

Customer	A large environment business company.
Industry	Private sector.
Project description	The client was a private sector body using an ageing, costly and increasingly unsupported set of personal productivity systems for its 12,500 users. The system integrator was engaged to investigate and subsequently to design, migrate and roll out an updated solution based upon Microsoft Office, Active Directory and Exchange software, implemented on a new physical infrastructure.
Solution	Microsoft Office, MS Exchange, Active Directory, Office Converter and Quest.
Team	Twelve staff directly involved in testing activities, of which five were onsite staff under the direction of the test manager/Scrum master.

Testing	• Enterprise wide integration testing of the infrastructure and packaged solutions
	• Design and build testing by the technical staff, but with additional testing expertise to help formalise and document the process
	• File migration testing using a checklist-driven approach
	• Exploratory application compatibility testing using parallel-run techniques in collaboration with key users
	• Back-end server performance testing using Microsoft's LoadSim and JetStress tools
Automation/ tool	Micro Focus test management tools used throughout, including requirements, test specification, test execution and defect management. Back-end server performance testing using Microsoft tools.

Customer	A large energy business company.
Industry	Utilities.
Project description	As part of a smart implementation programme, a number of strategic IT developments were delivered as a set of tactical solutions to support the programme. With the addition of new services and migration of data from legacy applications, it became apparent that some elements of the system architecture implemented under the original phase of the project needed to be refined to aid in the stability of the suite of applications and provide a base for the introduction of additional functionality and data volumes. The delivery was grouped into 10 releases comprising 23 major change requests and 24 major defect fixes.
Solution	Smart metering systems.
Team	The test team comprised five test analysts, two senior testers and a test manager/Scrum master.
Testing	As the third-party vendor performed limited system and integration testing, it was decided that the system integrator and client test team would conduct extensive system acceptance testing, system integration testing, field pilot testing and user acceptance testing to evaluate the risk.
Automation/ tool	T-Plan was the main test tool used but the strategy was to implement the Compuware suite.

Customer	Food and beverages.
Industry	MRD.
Project description	This delivery was for a major international company's international supply chain planning (ISCP) function and involved implementing the SAP Advanced Planning and Optimisation (APO) product globally. APO would support the enhanced analysis of predicted demand for the company's products and allow the creation of production schedules as part of the corresponding supply planning.
Solution	Design and implement the ISCP solution.
Team	System and UAT test team size was four including the test manager/Scrum master, in-house staff and client.
Testing	Component and component integration testing executed by multiple third-party suppliers. System and integration test, and production deployment test designed and executed by the SI; user acceptance test run jointly by the SI and the client.
Automation/ tool	Automated regression testing.

Customer	Retail business.
Industry	MRD.
Project description	Major Europe-based retail company had a requirement to re-platform its current software application suite.
Solution	Server upgrade and database upgrade.
Team	Test team size was seven including the test manager/Scrum master. One offshore tester.
Testing	Component and component integration testing executed by the third-party supplier; regression testing, performance testing, operability testing, system integration testing and user acceptance testing executed by the SI and the client.
Automation/ tool	LoadRunner and T-Plan were the main test tools used.

Customer	A large telecom company in Europe.
Industry	Telecommunications.
Project description	The client was a telecommunications company that had previously developed an advertising campaign management system based upon software packages by third parties. The original system had failed to deliver all the required functionality and benefits. This phase two project was designed to add the additional functionality, and so ensured that the original benefits of improved customer insight, customer care and increased value per customer were realised.
Solution	Campaign management solution.
Team	Test manager/Scrum master plus two professional testers, supported by five developers who also did testing.
Testing	System and integration testing.
Automation/ tool	Test management and automated regression testing.

Customer	Education based.
Industry	Public sector.
Project description	Web catalogue product maintenance system.
Solution	The project followed the standard Rational Unified Process (RUP) where there was one elaboration phase and four construction phases.
Team	The test team comprised four test engineers, a team leader and a test manager.
Testing	The main test activities were functional system testing, integration testing and regression testing.
Automation/ tool	IBM Rational Test Manager, ClearCase, ClearQuest and Rose.

Customer	Retail.
Industry	MRD.
Project description	Comprehensive reworking of the company's supply chain and warehouse processes for ordering and distributing goods, and associated updates to most of their IT systems to support these changes.
Solution	Supply chain and warehouse management.
Team	Test manager and 24 offshore testers.

Testing	• System testing
	• Integration testing
	• Performance testing
	• Operational testing
Automation/ tool	• TeamForge
	• Selenium

Customer	Travel business.
Industry	Travel and transport.
Project description	A core Salesforce solution created by developers with the customisation of the business processes being targeted to:
	• Improve the user experience
	• Create a single point and trusted source of data
	• Grow revenue and business
	• Reduce manual processes
	• Maximise and optimise the usage of the inventory
	• Allow for the decommissioning of the incumbent system
Solution	Salesforce cloud solution with customisation of the business.
Team	Test manager/Scrum master and three testers.
Testing	• Static testing
	• Functional testing
	• Regression testing
	• Accessibility testing
	• Role based testing
	• Compatibility testing
	• Risk based testing
Automation/ tool	• Micro Focus ALM and Jira for defect logging and tracking
	• Microsoft Excel for test case development, execution, tracking and reporting
	• TDD-based unit testing
	• Automated regression testing

Customer	Pensions and insurance business.
Industry	Insurance.
Project description	Pension.
Solution	Online web application for auditing suppliers who administer claimants' work assessments with integration assessment services portal.
Team	Test lead and two test automation engineers.
Testing	• Early integration testing (manual and automation) • Operational testing • System capacity testing (performance) • Accessibility testing
Automation/tool	• Jira • Selenium automated testing/GitLab CI • Postman • JMeter • Accessibility testing tools

Customer	Tax.
Industry	Public sector.
Project description	Testing of complex business rules validation on payloads, schemas and availability of systems accessed via APIs.
Solution	The solution created and managed APIs ensuring token validation, storage, publishing, data access, transformation policy management and security. The project had multiple Agile delivery teams with two test engineers and four developers.
Team	Two test automation engineers.
Testing	• Component testing • System testing • Regression testing • Smoke testing • Integration testing of CI
Automation/tool	• Jira, Confluence • JMeter, SoapUI • Post Man, Git, Jenkins, Wiremock, Mock JDBC, Virtualisation – WS02 EC2 virtual environment and virtualised server on D4D, XMLSpy and RabbitMQ

Customer	Leisure.
Industry	Travel and transport.
Project description	The existing pair of systems (partly on-premises and cloud) were transitioned from one supplier to another.
Solution	The solution involved upgrading and integrating the two systems into a single, powerful, stable system on cloud and delivering functional and performance enhancements.
Team	Two test leads.
Testing	• Multiple SIT cycles, user acceptance testing, functional testing, ELS, regression testing • End-to-end third-party acceptance testing, performance testing and OAT
Automation/ tool	Micro Focus ALM, Azure DevOps, BrowserStack, and WAVE API.

Customer	Energy business.
Industry	Geographic information systems (GISs) upgrade.
Project description	Implementing GE's Electric Office and integrating with IBM's Maximo Enterprise Asset Management System.
Solution	Consolidate and upgrade current small world to GE Electric Office Desktop and integrate Geospatial Analysis tool (Professional and Web), Mobile Enterprise solution.
Team	Two test leads.
Testing	• System testing • System integration testing • UAT • Regression testing • Performance testing
Automation/ tool	• Manual • Iterative • Automated

Customer	Pharmaceutical business.
Industry	Healthcare.
Project description	The solution encompassed a very complex workflow, including feasibility, plans, reviews, meetings, field work, inspections, quotes, payments and vesting (transfers of asset ownership).
Solution	• Azure Cloud Based Microsoft Dynamics CRM (customer relationship management) • Bespoke web portal • Dell Boomi ESB integration • SharePoint document repository
Team	Test lead and two testers.
Testing	• System testing • System integration testing • Accessibility compliance testing • Penetration testing
Automation/ tool	• Jira • Smartbear QA Complete

APPENDIX G: ISTQB TEST AUTOMATION ARCHITECTURE

The gTAA was developed by the ISTQB as part of the TAE certification syllabus. The gTAA provides an approach to implementing TASs for software projects. gTAA is a widely used and popular test automation architecture for designing, developing, implementing, and maintaining TASs. The gTAA-designed test automation architecture has a structure of four horizontal layers, which are test generation, test definition, test execution, and test adaptation. These layers are supported by the management layers, such as test management, project management, and configuration management (see Figure A.7).

Refer to the ISTQB TAE syllabus for further information on gTAA (https://www.istqb.org/downloads/category/48-advanced-level-test-automation-engineer-documents.html).

The TAE syllabus advises implementing the TAS in incremental steps (e.g. in sprints) for quick benefits with a recommendation of a proof of concept for test automation projects. The test automation project is generally managed as a software development project.

The gTAA is vendor independent and is often implemented by a set of tools and the components. The gTAA can be implemented by using any development approach, such as structured, object-oriented, service-oriented, or model-driven, as well as by any software technologies and tools. TAS is often implemented using off-the-shelf or open source tools but often needs additional SUT-specific additions and/or adaptations.

The test generation layer is generally used for designing test cases, and the test definition layer supports the definition and implementation of test suites or test cases. The test execution layer provides an execution tool to run the tests automatically along with a reporting component. The test adaptation layer helps the code to adapt the automated tests for various SUT components or interfaces. It provides different adaptors to connect with the SUT via APIs, protocols, services and so on.

The most common and popular test automation tools are suitable for implementing gTAA. The test automation planning and PoC stages are appropriate places to identify the components and layers of the gTAA required for your solution. The TAE syllabus provides suggestions about how to use a gTAA-based approach in an Agile and DevOps environment.

The gTAA is largely described as a monolith as it is composed of all its components in one architecture. The monolithic framework/application/architecture describes a single structure in which different components are combined into a single platform or a framework. Components can be business logic, database layer, API layer, or application integration and so on.

Figure A.7 The Generic Test Automation Framework ©International Software Testing
Qualifications Board (ISTQB), 2016

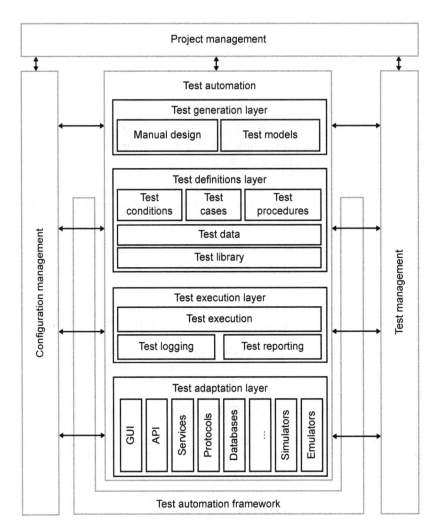

REFERENCES

Capgemini (2014) Managed Testing Services. https://www.capgemini.com/gb-en/resources/managed-testing-services/

Capgemini (2019) *World Quality Report 2019–20: Top Software Testing Trends for CIOs.* https://www.capgemini.com/research/world-quality-report-2019/

Capgemini (2020) *World Quality Report 20–21.* https://www.capgemini.com/gb-en/research/world-quality-report-wqr-20-21/

Microsoft (2020) Extension Methods (C# Programming Guide). https://docs.microsoft.com/en-us/dotnet/csharp/programming-guide/classes-and-structs/extension-methods

Prusak, L. (2010) What can't be measured. *Harvard Business Review*, 7 October. https://hbr.org/2010/10/what-cant-be-measured

FURTHER READING

Black, R. (2018) *Mobile Testing: An ASTQB-BCS Foundation Guide*. Swindon: BCS, The Chartered Institute for IT.

Fewster, M. (1999) *Software Test Automation*. New York: Addison-Wesley Professional.

Graham, D. and Fewster, M. (2012) *Experiences of Test Automation: Case Studies of Software Test Automation*, 1st edn. Upper Saddle River, NJ: Addison-Wesley Professional.

Hambling, B. (2019) *Software Testing: An ISTQB-BCS Certified Tester Foundation Guide*. Swindon: BCS, The Chartered Institute for IT.

Kaner, C., Bach, J. and Pettichord, B. (2002) *Lessons Learned in Software Testing: A Context-Driven Approach*. New York: John Wiley & Sons.

Lees, J. and Deluca, M. (2008) *Job Interviews: Top Answers to Tough Questions.* London: McGraw-Hill Education.

Paul, D., Cadle, J. and Yeates, D. (2020) *Business Analysis*, 4th edn. Swindon: BCS, The Chartered Institute for IT.

Perry, W. E. (2006) *Effective Methods for Software Testing*. New York: John Wiley & Sons.

Poundstone, W. (2005) *How Would You Move Mount Fuji? Microsoft's Cult of the Puzzle: How the World's Smartest Companies Select the Most Creative Thinkers*. New York: Little, Brown US.

GLOSSARY

.NET: This is a software platform/framework from Microsoft Corporation and is commonly called 'dot net'.

Abstraction: Displaying only the essential information and hiding background details.

Agile: Software development methodologies based on iterative and incremental development, where requirements and solutions evolve through collaboration between self-organising, cross-functional teams.

Annual licence: This allows the customer to use the licensed software for one year and after that the software will no longer work unless a new licence is purchased or the licence is renewed.

API: An application programming interface (API) is a particular set of rules ('code') and specifications that software programs can follow to communicate with each other. It serves as an interface between different software programs and facilitates their interaction.

Azure DevOps: A Microsoft toolset, which is used to store product backlog items, sprint information and code.

Business component: A reusable unit that performs a specific task in a business process. You can use a component in multiple business process tests and flows. When you modify a component or its steps, all business process tests or flows containing that component reflect that modification.

Business requirements document: Defines the functional and non-functional specifications for the project or product.

CapEx: Capital expenditures (CapEx) are funds used to acquire, upgrade and maintain assets such as infrastructure, plants, buildings, technology and equipment.

Change control: A procedure to ensure that all changes are controlled, including the submission, analysis, decision making, approval, implementation, documentation and post implementation changes/business as usual (BAU) changes.

Class: In object-oriented programming, a class is an extensible program for creating objects.

Cloud: Cloud computing is storing and accessing data and programs over the internet instead of on your computer's hard drive. The PCMag Encyclopedia (https://www.pcmag.com/encyclopedia/a) defines it succinctly as 'hardware and software services from a provider on the internet').

Code review: A systematic check of other's code for mistakes, based on standards and guidelines.

Coding: The process of using a programming language to create a set of instructions for a computer to behave as would be expected.

Commercial off-the-shelf: Equipment or software, as applicable, that is readily available to the public from a third party rather than custom-made or bespoke.

Common libraries: A collection of useful libraries that are used for automated testing. They are a source of information referenced and used to fulfil defined common activity.

Component design architecture document: This provides the detailed architecture for a component. It describes the services a component provides, its sub-components and the supporting architectural views for a component, for example application, data and security architectures.

Concurrent licence: This licence is based on the number of simultaneous or concurrent users accessing the software at a single point of time.

Configuration management: This is the tracking and controlling of changes in the software including version control. It is essential once the software has been developed and released by multiple teams or multiple environments across a wide range of end-user communities.

Continuous integration and continuous delivery (CI/CD): Also known as continuous deployment. CI is a set of practices performed as developers are writing code, and CD is a set of practices performed after the code is completed, that is, deployment.

Cucumber: A software tool that supports behaviour-driven development (BDD).

Cybersecurity assessment: This provides an insight into the security aspects and risks associated with the implementation of the IT solution and is used subsequently to show that security has been adequately considered.

Debugging: The process of detecting bugs and errors in a software code.

Deliverable: Anything provided or to be provided by the supplier under the agreement as a result of the services, which could include work items, software, hardware, documentation, reports, drawings, calculations, recommendations and conclusions.

Delivery: The conformity of a particular work item with the relevant 'definition of done' and acceptance criteria; deliver and delivered shall be construed accordingly.

Deployment: Software deployment is the process of making the application available on a target place, for example test server, production environment, mobile device.

Detailed design document: This defines the design of the physical architecture of the selected logical option, server types, network architecture, storage requirements and interfaces. This also includes any network design.

DevOps: (Development operations) The close relationship between the developers of applications and the people who test and deploy them. DevOps is said to be 'the intersection of software engineering, quality assurance and operations. It focuses on improving collaboration among teams' – according to the PCMag Encyclopedia.

Done: The criteria that need to be met in order for the product owner to accept a product backlog item as being complete.

Early life support (ELS): The services to be provided by the supplier after the release/ delivery.

Enhancement: Any modification, update or new release of the software that corrects faults, adds functionality or otherwise amends or updates.

Environment: A subset of the IT infrastructure that is used for a particular purpose (for example production environment, test environment and development environment).

Epic: A collection of features or user stories.

Exception handling: This is the process in a program responding to exceptions or unexpected behaviour.

Feature: A collection of user stories.

Fixed price (FP) contract: This is a type of contract where the payment amount is fixed and is not dependent on resources used or time expended.

Full system architecture document: Defines the generic structure of the solution, providing the logical and physical information architecture for the selected option, including all related components and interfaces, capacity and service continuity designs.

General Data Protection Regulation (GDPR): Regulation (EU) 2016/679 of the European Parliament and of the Council of 27 April 2016 on the protection of natural persons with regard to the processing of personal data and repealing directive 95/46/EC (General Data Protection Regulation) OJ L 119/1, 4.5.2016. The GDPR is a European Union compliance regulation that went into effect in May 2018.

GUI object: An object has state (data) and behaviour (code) and is an abstract data type with the addition of polymorphism and inheritance. A GUI (graphical user interface) object is the same, representing GUI elements such as a text box, a button or a check box.

High-level design (HLD): Defines the generic structure of the solution, providing the logical (and physical) information architecture for the selected option, including all related components and interfaces, capacity and service continuity designs.

I18N: The abbreviation of the word 'internationalisation', being the letter 'I' followed by 18 more letters, followed by 'N'.

Inception: The early activity required to do just enough work to get the team going in the right direction. This can include, but is not limited to: initial requirements modelling, initial architecture modelling, initial planning, initial organisational modelling.

Infrastructure as a service (IaaS): According to the PCMag Encyclopedia, 'IaaS is a cloud computing service that provides a basic computing platform, typically the hardware and virtual machine (VM) infrastructure (no operating system) or the hardware and an operating system'.

Inheritance: The ability of one class of objects to inherit properties from a higher class.

ISTQB: ISTQB (International Software Testing Qualifications Board) was founded in November 2002 and is a not-for-profit association. ISTQB has defined the 'ISTQB Certified Tester' scheme that has become the worldwide leader in the certification of competences in software testing.

Iteration: A short-time boxed period, typically 2 to 4 weeks in duration, during which time the feature team(s) completes as much of the work planned for the iteration as it is reasonably able to do.

IT service continuity (ITSC): The process of ensuring that identified IT services will be available during abnormal situations. It typically involves a detailed assessment of the business risk of key IT services being lost, and then identifies countermeasures and plans to prevent, or recover from, identified contingencies.

Java: An object-oriented programming language from Oracle Corporation, platform independent.

L10N: An abbreviation of the word 'localisation', which is the letter 'l' followed by 10 more letters, followed by 'n'.

Lean: Lean software development adapted principles from the Toyota Lean production system and is largely used by the Agile community. It focuses on optimising efficiency and minimising the waste of software applications during their design and development. The value, value stream, flow, pull, and perfection are considered as the five key Lean principles.

Low-level design (LLD): This defines the low-level design of the physical architecture of the selected logical option, server types, network architecture, storage requirements and interfaces.

Minimum viable/valuable product (MVP): An early version of a working product with just enough features to be usable by early users who can then provide feedback for future product development.

Non-functional requirement (NFR): A requirement that does not relate to functionality, but to attributes such as reliability, efficiency, usability, maintainability and portability.

Object-oriented programming (OOP): A programming structure wherein the data and their methods are defined as self-contained entities called 'objects'.

Object repository: A collection of objects and properties taken from the system under test (SUT), so that a testing tool can recognise the corresponding elements. Record and playback gathers the objects and their properties.

OpEx: An operating expense (OpEx) is an expense required for the day-to-day functioning of a business.

Personal identifiable information (PII): Any data that could potentially be used to identify a particular person. Examples include a full name, National Insurance number, driver's licence number, bank account number, passport number or email address.

Physical design document: This defines the design of the physical architecture of the selected logical option, server types, network architecture, storage requirements and interfaces. This also includes any network design.

Platform as a service (PaaS): A cloud computing service that provides a comprehensive computing environment. 'PaaS includes the hardware, operating system, database and other necessary software for the execution of applications. It may include a complete development environment as well' (according to the PCMag Encyclopedia).

Polymorphism: The ability of an object to take on many forms based on the object that it is sent to.

Portability: The degree of compatibility and ease of moving applications, data, source code and so on from one operating system to another or one database to another or one cloud to another or one platform to another.

PRINCE2: Projects in Controlled Environments 2 is a structured project management approach and practitioner certification programme.

Product: A combination of features that supports a particular business process or group of processes.

Product backlog: A prioritised list of items, containing short descriptions of all functionality desired in the product, and associated business outcomes as defined. A product backlog comprises: (i) the prioritised set of work items; (ii) the estimated amount of story points required to deliver each work item; (iii) the work items with the status of done; and (iv) acceptance criteria, to be delivered as part of the project, as amended from time to time by the product owner in accordance with this schedule.

Product manager: A person that provides long-term ownership and vision for the product.

Product owner: The main representative concerning the scope and requirements of a product and accountable for updating the product backlog.

Project: The overall delivery of the services described in a particular package.

Project manager: A person who is responsible for the delivery of large IT projects, acquiring and utilising the necessary resources and skills, within the agreed parameters of cost, timescales and quality.

Proof of concept (PoC): The evidence that a proposed product concept is viable and capable of solving an organisation's problem.

Proof of technology (PoT): The evidence that a proposed product technology is a viable solution that can solve an organisation's problem.

Quality assurance (QA): The process of ensuring that the quality of a product, service or process will provide its intended value.

RACI log: Refers to a responsible, accountable, consulted and informed matrix.

RAID log: Refers to a risks, actions, issues and decisions matrix.

Record and playback: A type of automated testing where the automation tool records the user behaviour or activity and plays this back.

Recovery point objective (RPO): The longest period of time in which newly entered data can be lost in the event of a major IT failure. The RPO determines how often backups must be performed.

Recovery scenario: This contains generic functions to handle exceptions at runtime.

Recovery time objective (RTO): The amount of time a computer system or application can stop functioning before it is considered unacceptable to the organisation. The RTO is used to determine the types of backup and disaster recovery plans that should be implemented.

Release: A collection of hardware, software, documentation, processes or other components required to implement the solution. The contents of each release are managed, tested and deployed as a single entity.

Scrum: Scrum is based on a 'Sprint', which is typically a predefined period for delivering a working part of the system. Each Sprint starts with a planning session that includes the customer (product owner), the facilitator (Scrum master) and the cross-functional team.

Scrum master: The facilitator in a Scrum software project.

Seat licence: 'A software licence based on the number of users who have access to the software. For example, a 100-user licence based on users means that up to 100 specifically named users have access to the program' (according to PCMag Encyclopedia).

Security architecture document: This provides insight into the security aspects and risks associated with the implementation of the solution and used subsequently to show that security has been adequately considered. It highlights the threats and vulnerabilities and investigates countermeasures to provide assurance to the business that adequate security has been implemented.

Smoke test: Validates the key features and basic functionalities of a software program or solution to ensure that it is fit for further detailed testing. Smoke test was originally used to find leaks in containers and pipes using smoke. The term was introduced to software testing with reference to testing a software application for the first time.

Software as a service (SaaS): A rented software service. Instead of buying applications, with this you pay a subscription, that is, when that expires, the software is no longer valid.

Software development life cycle (SDLC): The sequence of events in the development of an information system.

Software test life cycle (STLC): The sequence of events in the testing of an information system.

Solution: The solution delivered by the supplier under this project package, which may comprise more than one product.

Solution architecture design document: This defines the generic structure of the solution, providing the logical and physical information architecture for the selected option, including all related components and interfaces, capacity and service continuity designs.

Sprint: A time-boxed period during which the team delivers the Sprint backlog.

Sprint backlog: The list of product backlog items (PBIs) that the product team commit to deliver in the Sprint.

Sprint goal: A short statement summarising the principal objective of the Sprint.

Sprint planning: The meeting at the start of a Sprint where the product owner states the Sprint goal, presents the PBIs that they would like to have completed in the next Sprint and agrees the Sprint backlog.

Sprint retrospective: The meeting at the end of a Sprint where the product team review the delivery methodology and agree changes to be implemented in the next Sprint.

Sprint review: The meeting at the end of a Sprint where the team present the results and output of the Sprint to the product owner and stakeholders, and adaptions to the product backlog are discussed and agreed, if needed.

Stakeholder: Those who have some interest in the products or projects. Any individual who may be affected by a business decision.

Static code analysis: Implies the usage of tools that scan the source code to find out if it contains any of the known formal defect patterns.

Story: A description of a particular behaviour of the product agreed upon by the developers, the product owner and the architecture owner as part of the Sprint planning process.

Subcontractor: Any party engaged by the supplier in the provision of the services.

Subject matter expert (SME): An individual who hold knowledge of a product or services.

System(s) software: Those programs and software, including documentation and materials, that perform tasks basic to the functioning of the computer hardware, or that are required to operate the applications, or otherwise support the provision of services by the supplier. Systems software includes operating software, systems utilities and any other software not designated as applications.

Test environment: A controlled environment used to test configuration items, builds, IT services and processes.

Test plan: This may comprise separate documents for each test phase in line with the test strategy. This includes all the information necessary to plan and control the test effort for the solution development phase. It describes the approach to the testing of the artefacts and is the top-level plan generated and used by the managers to direct the test effort.

Test pyramid: According to ISTQB Glossary, test pyramid is a graphical model representing the relationship of the amount of testing per level, with more testing at the bottom than at the top. Mike Cohn, one of the founders of the Scrum Alliance, came up with the test pyramid concept in his book *Succeeding with Agile*. It is a visual model of different layers of testing and how much testing is to be done on each layer. Mike Cohn's original test pyramid consists of three layers including unit tests, service tests and user interface tests (bottom to top). Unit tests form the base of the test pyramid, that should be frequent, and run fast. Integration tests are the middle tier of the pyramid. Test pyramid expects that more automated testing is done through unit tests than GUI-based testing. The pyramid provides the advantages of avoiding many of the complexities of dealing with UI frameworks.

Test script: An automation test script is a set of instructions that will be performed on the system under test (SUT) to validate that it works as expected. Good test scripts are small, isolated, maintainable and easy to understand with swift execution of steps. They are independent of each other and run with good exception handling.

Test strategy: This defines the strategy for the assurance and testing effort required for the project/product. It provides a summary of the test stages and responsibilities specific to the project. Test objectives, scope and resources are detailed within this strategy.

Testing as a service (TaaS): A model in which IT solution testing activities associated with an organisation's activities are performed by a supplier rather than in-house employees.

Time and materials (T&M): A standard contract for product development or any other piece of work in which the customer agrees to pay the supplier based upon the time spent and materials used.

Traffic light/RAG: A RAG rating system indicates the status of a variable using red, amber or green traffic lights.

User story: A description of a particular behaviour of the product in a few sentences.

Validation: Typically involves actual testing and takes place after verifications are completed.

Verification: Typically involves reviews and meetings to evaluate documents, plans, code, requirements, and specifications. This can be done with checklists, issues lists, walkthroughs, and inspection meetings.

Work item: An item in the product backlog, including but not limited to enhancements, epics, features, user stories or tasks.

INDEX

www.ingramcontent.com/pod-product-compliance
Lightning Source LLC
Chambersburg PA
CBHW060529060326
40690CB00017B/3432